It's another Quality Book from CGP

This book is for anyone doing AQA Modular GCSE Mathematics at Foundation Level.

Whatever subject you're doing it's the same old story — there are lots of facts and you've just got to learn them. KS4 Maths is no different.

Happily this CGP book gives you all that important information as clearly and concisely as possible.

It's also got some daft bits in to try and make the whole experience at least vaguely entertaining for you.

What CGP is all about

Our sole aim here at CGP is to produce the highest quality books — carefully written, immaculately presented and dangerously close to being funny.

Then we work our socks off to get them out to you — at the cheapest possible prices.

Contents

Published by CGP

Written by Richard Parsons

Updated by: Helena Hayes, David Ryan, Jane Towle, Janet West, Dawn Wright

Proofreading by: Margaret Darlington, Julie Wakeling

ISBN: 978 1 84146 543 2

Groovy website: www.cgpbooks.co.uk
Printed by Elanders Ltd, Newcastle upon Tyne.
Jolly bits of clipart from CorelDRAW®

Calculating Tips

Ah the glorious world of GCSE Maths. OK — maybe it's more like a wiffy sock at times, but do it you must I'm afraid. AQA Modular Maths is split into 3 Units, each with their own exam — yuk. Thankfully there's some maths and nifty exam tricks you only have to learn once, and could get you marks in all of 'em.

BODMAS

Brackets, Other, Division, Multiplication, Addition, Subtraction

BODMAS tells you the ORDER in which these operations should be done:
Work out Brackets first, then Other things like squaring, then Divide / Multiply groups of numbers before Adding or Subtracting them.

This set of rules works really well for simple cases, so remember the word BODMAS.

Example: A mysterious quantity T, is given by: $T = (P - 7)^2 + 4R/Q$
Find the value of T when P = 4, Q = -2 and R = 3

Write down the formula: $T = (P - 7)^2 + 4R/Q$

Put the numbers in: $T = (4 - 7)^2 + 4\times3/{-2}$

Then work it out in stages:

$= (-3)^2 + 4\times3/{-2}$

$= 9 + 4\times3/{-2}$

$= 9 + -6$

$= 9 - 6 = \underline{3}$

Note BODMAS in operation: Brackets worked out first, then squared. Multiplications and divisions done before finally adding and subtracting.

Make Sure You Know What Your Answer Means

It's taken 2 minutes of frenzied button pressing and finally your calculator screen looks like this.

3.6

Before you merrily jot down 3.6 as your answer, think what it means — 3.6 what? Pipers piping? It sounds silly, but it can lose you easy marks in the exam. E.g. If you're answering a money question, £3.6 won't get you any marks — you need to write £3.60.

And now for something completely similar...

Make Sure You Include the Units

This is probably pretty obvious, but it's an easy thing to forget in the exam so...

ALWAYS CHECK WHETHER YOUR ANSWER NEEDS A UNIT

The question might already say what units the answer needs to be in, or you might need to work it out for yourself. This isn't too difficult.

For example, if it's a question about area, and the measurements are given in cm, then your answer will need to be given as cm^2.

Or, if it's a question about volume, and the measurements are given in m, then your answer will need to be given as m^3.

Unit to include the units, unit wit...

This might all seem like really boring obvious stuff, but you'd be amazed how many people forget about these little things in the exam. But you won't will you. You'll read over this page again, just to make sure you've got it all.

Calculating Tips

Always Check Your Answer

It's always a good idea to check your answers by working backwards through your calculations.

That way you can pick up any silly mistakes you might have made.

Example: If $y = 2x^2 + 3$, find the value of y when $x = 3$.

Answer: $y = 2 \times (3)^2 + 3$
$= 2 \times 9 + 3$
$= 18 + 3$
$= 21$

Check: $21 - 3 = 18$
$18 \div 2 = 9$
$\sqrt{9} = 3$ — Hurray — you've got the same number you started with.

Don't Do Any Rounding Until the End of the Calculation

The first few pages of this book are all about rounding. But before you get all giddy from your mad new skillz and start rounding left, right and centre, a word of warning. Don't.

Example: Find the value of $100\sqrt{2}$ to 1 d.p.

Wrong Wrong Wrong

$100\sqrt{2} = 100 \times 1.4$
$= 140.0$ (1 d.p.)

This poor soul rounded √2 to 1 d.p. halfway through the calculation. What a numpty — look how wrong their answer is.

Right Right Right

$100\sqrt{2} = 100 \times 1.41421356...$ ← No rounding here!
$= 141.4$ (1 d.p.)

Much better to store √2 in your calculator's memory (see p. 7), and then just round at the end. Hooray for rounding at the end.

Don't Be Scared of Wordy Questions

About a third of the marks on your exam come from wordy, real-life questions. For these you don't just have to do the maths, you've got to work out what the question's asking you to do first. Relax and work through them step by step.

1) READ the question carefully. Work out what bit of maths you need to answer it.
2) Underline the INFORMATION YOU NEED to answer the question — you might not have to use all the numbers they give you.
3) Write out the question IN MATHS and answer it, showing all your working clearly.

Example:

> A return car journey from Carlisle to Manchester uses 4/7 of a tank of petrol.
> It costs £56 for a full tank of petrol.
> How much does the journey cost?

It doesn't matter where they're driving from and to.

1) The "4/7" tells you this is a fractions question.
2) You need £56 (the cost of a full tank) and 4/7 (the fraction of the tank used).
3) You want to know 4/7 of £56, so in maths:

£56 × 4/7 = £32 ← Don't forget the units in your final answer — this is a question about cost in pounds, so the units will be £.

So close to starting now — excitement building, heart racing...

Learn this stuff — it can really help you rack up marks whichever exam you're doing. Right, on with the rest of the show. Laaaaaaaadies and gentlemeeeeeeen — I give you, the one, the only — AQA Maths G C S Eeee.

Rounding Off

When you have decimal numbers you might have to round them off to the nearest whole number. The trouble is, they could also ask you to round them off to one, two or three decimal places. This isn't too bad but you do have to learn some rules for it:

Basic Method

1) Identify the position of the LAST DIGIT.

2) Then look at the next digit to the right – called the DECIDER.

3) If the DECIDER is 5 or more, then ROUND-UP the LAST DIGIT.
If the DECIDER is 4 or less, then leave the LAST DIGIT as it is.

EXAMPLE: What is 7.35 to 1 Decimal Place?

7.35 = 7.4

LAST DIGIT to be written (because we're rounding to 1 Decimal Place)

DECIDER

The LAST DIGIT ROUNDS UP to 4 because the DECIDER is 5 or more

Decimal Places (D.P.)

1) To round off to ONE DECIMAL PLACE, the LAST DIGIT will be the one just after the decimal point.
2) There must be NO MORE DIGITS after the LAST DIGIT (not even zeros).

EXAMPLES

Round off 2.34 to 1 decimal place. ANSWER: 2.3
Round off 4.57 to 1 decimal place. ANSWER: 4.6
Round off 2.084 to 2 decimal places. ANSWER: 2.08
Round off 2.3465 to 3 decimal places. ANSWER: 2.347

Rounding off decimal places — I totally dig it...

Personally, I quite like the rounding off methods on this page. They make it easier to explain to my numbers why I've rounded them up or down — I just blame the rules. They take it less personally then.

1) Round each of these numbers off to 1, 2 and 3 decimal places:
a) 3.2456 b) 1.7801 c) 2.3115 d) 0.4674 e) 9.7646

2) Round these off to the nearest whole number:
a) 3.4 b) 5.2 c) 1.84 d) 6.9 e) 3.26

Rounding Off

If you thought that it was only numbers with decimal places that could be rounded off — then you're wrong. So very, very wrong. Whole numbers can be rounded off too. Yay.

Rounding Whole Numbers

The easiest ways to round off a number are:

1) 'To the nearest WHOLE NUMBER'
2) 'To the nearest TEN'
3) 'To the nearest HUNDRED'
4) 'To the nearest THOUSAND'

This isn't difficult so long as you remember the 2 RULES:

> 1) The number always lies between 2 POSSIBLE ANSWERS, Just choose the one it's NEAREST TO.
> 2) If the number is exactly in the MIDDLE, then ROUND IT UP.

EXAMPLES:

1) Give 231 to the nearest TEN.

ANSWER: 231 is between 230 and 240, but it is nearer to 230

2) Give 145 to the nearest HUNDRED.

ANSWER: 145 is between 100 and 200, but it is nearer to 100

3) Round 45.7 to the nearest WHOLE NUMBER.

ANSWER: 45.7 is between 45 and 46, but it is nearer to 46

4) Round 4500 to the nearest THOUSAND.

ANSWER: 4500 is between 4000 and 5000. In fact it is exactly halfway between them. So we ROUND IT UP (see Rule 2 above) to 5000

Significant Figures

> 1) The MORE SIGNIFICANT FIGURES a number has, the MORE ACCURATE it is.
> 2) The NUMBER OF SIGNIFICANT FIGURES is just HOW MANY DIGITS the number has at the front THAT ARE NOT ZERO.

EXAMPLES:

234 has 3 significant figures
9810 has 3 sig fig
230 has 2 sig fig
4000 has 1 sig fig
900 has 1 sig fig
2.8 has 2 sig fig

Exam cheat wish #23 — all marks are rounded up to the nearest 100%...

Learn the 2 Rules for Rounding Whole Numbers and the 2 Extra Rules about Significant Figures. Simples.

1) Round these off to the nearest 10:
   a) 453 b) 682 c) 46.2 d) 98 e) 14
2) Round these numbers to one significant figure:
   a) 352 b) 465 c) 12.38 d) 0.03567
3) Round these numbers off to the nearest hundred: a) 2865 b) 450 c) 123

Calculator Buttons

In your Unit 1 Exam you'll be allowed to use a calculator, so it's really important you know how your calculator works. But remember — not all calculators are the same.

The FRACTION BUTTON [$a\frac{b}{c}$]

You do need to know how to deal with fractions without your calculator.
But when you're allowed to use it, you definitely should do...

1) TO ENTER A NORMAL FRACTION like $\frac{1}{4}$

Just press: [1] [$a\frac{b}{c}$] [4]

DISCLAIMER
The buttons on this page are merely an artist's interpretation of the buttons featured on an actual calculator — pressing them will have little or no effect.

2) TO ENTER A MIXED FRACTION like $1\frac{3}{5}$

Just press: [1] [$a\frac{b}{c}$] [3] [$a\frac{b}{c}$] [5]

3) TO DO A REGULAR CALCULATION such as $\frac{1}{5} \times \frac{3}{4}$

Just press: [1] [$a\frac{b}{c}$] [5] [×] [3] [$a\frac{b}{c}$] [4] [=]

It's the same for adding or subtracting fractions e.g. $\frac{1}{5} + \frac{3}{4}$
Just press: [1] [$a\frac{b}{c}$] [5] [+] [3] [$a\frac{b}{c}$] [4] [=]

4) TO REDUCE A FRACTION TO ITS LOWEST TERMS

Just enter it and then press [=].
e.g., [9] [$a\frac{b}{c}$] [12] [=] → 3⌟4 $= \frac{3}{4}$

5) TO CONVERT BETWEEN MIXED AND TOP HEAVY FRACTIONS

Just press [SHIFT] [$a\frac{b}{c}$]   e.g. to give $2\frac{3}{8}$ as a top heavy fraction:
Press: [2] [$a\frac{b}{c}$] [3] [$a\frac{b}{c}$] [8] [=] [SHIFT] [$a\frac{b}{c}$] which gives an answer of $\frac{19}{8}$.

The Powers Button — [x^y] or [^]

It's used for working out powers of numbers quickly. For example to find 7^5, instead of pressing $7 \times 7 \times 7 \times 7 \times 7$, you should just press [7] [x^y] [5] [=].

(See P.41 for more on powers)

Calculator Buttons

The BRACKETS BUTTONS (and)

1) These are really important when you want to work out even a simple thing like $\frac{23+45}{64\times 3}$.
2) You can't just press 23 + 45 ÷ 64 × 3 = — it will be completely wrong.
   The calculator will think you mean $23+\frac{45}{64}\times 3$ because the calculator will do the division and multiplication BEFORE it does the addition. This is because it uses BODMAS (see p.2).
3) The secret is to OVERRIDE the automatic BODMAS order of operations using the BRACKETS BUTTONS. Brackets are the ultimate priority in BODMAS, which means anything in brackets is worked out before anything else happens to it.

> So all you have to do is:
> 1) Write a couple of pairs of brackets into the expression: $\frac{(23+45)}{(64\times 3)}$
> 2) Then just type it as it's written:
>
> (23 + 45) ÷ (64 × 3) =

4) You might think it's difficult to know where to put the brackets in.
   It's not that difficult, you just put them in pairs around each group of numbers.
5) It's OK to have brackets within other brackets too, e.g. (4 + (5÷2)).
   As a rule, you can't cause trouble by putting too many brackets in,

SO LONG AS THEY ALWAYS GO IN PAIRS.

The MEMORY BUTTONS — STO Store, RCL Recall

These are really useful for keeping a number you've just calculated, so you can use it again shortly afterwards.

EXAMPLE: Find $\frac{840}{15+5^3}$ — just work out the bottom line first and stick it in the memory.

So press 15 + 5 x^3 = and then STO M to keep the result of the bottom line in the memory.
Then you simply press 840 ÷ RCL M =, and the answer is 6.

> The memory buttons might work a bit differently on your calculator. Note, if your calculator has an 'Ans' button, you can do the same thing as above using that instead — the Ans button gives you the result you got when you last pressed the '=' button.

Learn these two pages, store, then recall...

Remember to practise everything you've just learnt on these two pages on your own calculator.

1) Work out $\frac{3}{4}\times\frac{1}{2}$.
2) Write down what buttons you would press to work this out in one go: $\frac{23.3+35.8}{36\times 26.5}$

Fractions

The next few pages cover fractions and percentages, but don't worry, they're quite similar. In fact fractions, percentages and decimals are just different ways of saying the same thing (see page 43). But for now make sure you can do everything explained here both with and without your calculator.

Equivalent Fractions

Equivalent fractions are fractions that are equal in value, even though they look different.

Starting with any fraction you like, you can make up a list of equivalent fractions by simply MULTIPLYING top and bottom by the SAME NUMBER each time:

$$\frac{1}{2} = \frac{3}{6} \quad (\times 3 \text{ top and bottom})$$

$$\frac{3}{4} = \frac{15}{20} \quad (\times 5 \text{ top and bottom})$$

$$\frac{1}{5} = \frac{100}{500} \quad (\times 100 \text{ top and bottom})$$

Cancelling Down

Going the other way, you will sometimes need to simplify a fraction by 'cancelling down' — which only means DIVIDING top and bottom by the SAME NUMBER:

$$\frac{3}{15} = \frac{1}{5} \quad (\div 3 \text{ top and bottom})$$

$$\frac{22}{33} = \frac{2}{3} \quad (\div 11 \text{ top and bottom})$$

Finding a Fraction of Something

Multiply the 'something' by the TOP of the fraction, then divide it by the BOTTOM:

$$\frac{9}{20} \text{ of } £360 = \{(9) \times £360\} \div (20) = \frac{£3240}{20} = £162$$

$$\text{or: } \frac{9}{20} \text{ of } £360 = \frac{9}{1} \times £360 \times \frac{1}{20} = £162$$

Let's be honest — all of these are pretty easy with a calculator.

It's a good idea to learn the method behind them though. Especially for Unit 2, when you can't use a calculator.

No fractions were harmed in the making of this page...

Sadly you've got to be able to mix and match all the stuff you're learning. Horrific, I know. Now you know how to cancel down and find fractions try the questions below.

1) Reduce these fractions to their simplest form: a) $\frac{30}{36}$ b) $\frac{18}{27}$ c) $\frac{45}{66}$

2) a) Find $\frac{2}{5}$ of 550. b) What's $\frac{7}{8}$ of £2?

Percentages

These two pages are all about percentages in real-life situations. It's a thinly veiled attempt by examiners to make maths more 'relevant'. So let's humour them — it's bound to be in the exam.

Discounts, VAT, Interest, Increase, etc.

MOST percentage questions are like this: **Work out 'something %' of 'something else'.**

E.g. "Find 20% of £60."

This is the method to use:

1) WRITE IT DOWN: Find 20% of £60.

2) TRANSLATE INTO MATHS: $\frac{20}{100} \times 60$

3) WORK IT OUT: 20 ÷ 100 × 60 = £12

Two Important Details:

Make sure you remember them!

1) 'Per cent' means 'out of 100'
so 20% means '20 out of 100' = 20 ÷ 100 = $\frac{20}{100}$

(That's how you work it out in the method shown above)

2) 'OF' means '×'
In maths, the word 'of' can always be replaced with '×' for working out the answer.
(as shown in the above method)

Percentages Without a Calculator

For Unit 1, you're allowed to use a calculator, but this method is still really useful to know (and it's really important when you come to do percentages in Unit 2). The way to do it without a calculator is to find 10% or 1% of the number, then use those values to get to the percentage you need.

E.g. "Find 40% of £30."

10% of £30 $= \frac{1}{10} \times$ £30 = £30 ÷ 10 = £3

So 40% of £30 = 4 × £3 = £12

10% $= \frac{10}{100} = \frac{1}{10}$
...so to find 10%, divide by 10.

E.g. "Find 6% of £500."

1% of £500 $= \frac{1}{100} \times$ £500 = £500 ÷ 100 = £5

So 6% of £500 = 6 × £5 = £30

1% $= \frac{1}{100}$
...to find 1%, divide by 100.

Percentages

Important Example — % Discount

E.g. A radio is priced at £8.50 but there is a discount of 20% available.
FIND THE REDUCED PRICE OF THE RADIO.

ANSWER:
First find 20% of £8.50 using the method from the last page:

1) 20% of £8.50
2) $\frac{20}{100} \times 8.5$
3) 20 ÷ 100 × 8.5 = 1.7 = £1.70

It's money, so 1.7 on the calculator display is £1.70.

This is the DISCOUNT so we take it away to get the final answer: £8.50 – £1.70 = £6.80

E.g. A shirt normally costs £15, but it has been discounted by 8%.
WITHOUT using your calculator, FIND THE REDUCED PRICE OF THE SHIRT.

ANSWER:
1% of £15: $£15 \times \frac{1}{100} = £15 \div 100 = £0.15 = 15p$

You might have guessed this bit — 1% of any amount in £ is that amount in pence.

So 8% of £15 = 15p × 8 = (10p × 8) + (5p × 8) = 80p + 40p = 120p = £1.20.

This is the DISCOUNT so take it away to get the final answer: £15.00 – £1.20 = £13.80

Important Example — adding VAT

E.g. A plumber's bill for fixing a small leak is £98 + VAT.
The VAT is charged at 17.5%. **WORK OUT THE TOTAL BILL.**

Answer: First find 17.5% of £98 using the standard method:

1) 17.5% of £98
2) $\frac{17.5}{100} \times 98$
3) 17.5 ÷ 100 × 98 = 17.15 = £17.15

This £17.15 is the VAT which then has to be ADDED to the £98 to give the FINAL BILL:
£98 + £17.15 = £115.15

Reality TV show error #52 — it's impossible to give 110%......

Always use the three part method for answering percentage questions. And if your question is like the ones above — don't forget you might need to add or subtract your answer to get your final answer.

1) A bank charges interest at 12% per year. If £1000 is borrowed for one year, how much interest will be charged?

Percentages

In an Exam you might be asked to compare two numbers using percentages. Wowsers. And by some happy coincidence that's what this page is all about.

Comparing Numbers using Percentages

This is the other common type of percentage question.

Give 'one number' AS A PERCENTAGE OF 'another number'

For example, "Express £2 as a percentage of £20." This is the method to use:

The FDP Method:

Fraction		Decimal		Percentage
$\frac{2}{20}$	$2 \div 20$ →	0.1	$\times$ 100 →	10%
First make a FRACTION out of the 2 numbers, always with the smaller number on top.		DIVIDE them to get a DECIMAL.		Then multiply by 100 to get a PERCENTAGE.

Two Important Examples

1) "A shopkeeper buys pens at 8p each and sells them for 10p each. What is his profit AS A PERCENTAGE?"

Answer: The two numbers we want to compare are the PROFIT (which is 2p) with the ORIGINAL cost (which is 8p). We then apply the FDP method:

Fraction → Decimal → Percentage:

$\frac{2}{8}$ → 0.25 → 25%

so the shopkeeper makes a 25% profit on the pens.

2) "In a sale, a tennis racket is reduced in price from £60 to £48. What PERCENTAGE REDUCTION is this?"

Answer: The two numbers we want to compare are the REDUCTION (which is £12 and the ORIGINAL VALUE (which is £60). We then apply the FDP method:

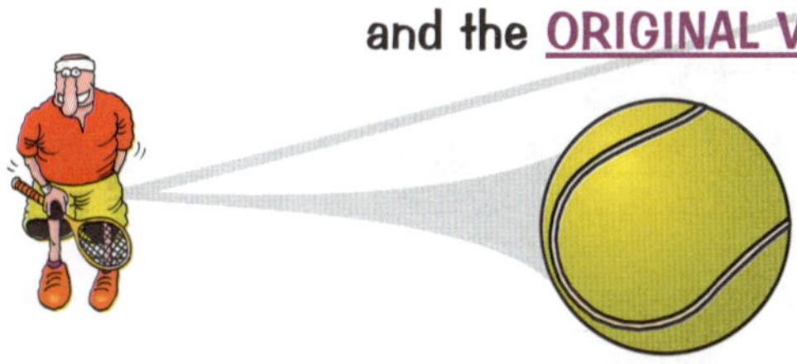

Fraction → Decimal → Percentage:

$\frac{12}{60}$ → 0.2 → 20%

63% of all statistics are made up...

So, the 'ole FDP method comes in handy here too. Time to check what you've learnt from this page...

1) A house increases in value from £140 000 to £182 000. What is the increase in value of the house as a percentage?

Ratios

There are lots of Exam questions which at first sight seem completely different but in fact they can all be done using the GOLDEN RULE...

DIVIDE FOR ONE, THEN TIMES FOR ALL

Example 1:

"5 pints of Milk cost £1.30. How much will 3 pints cost?"

The GOLDEN RULE says: **DIVIDE FOR ONE, THEN TIMES FOR ALL**

which means: Divide the price by 5 to find how much FOR ONE PINT, then multiply by 3 to find how much FOR 3 PINTS.

So... £1.30 ÷ 5 = 0.26 = 26p (for 1 pint)
×3 = 78p (for 3 pints)

Continuing the dairy theme, Example 2 on page 45 is worth learning for this Unit as well...

Example 2:

"The following recipe is for 'Froggatt's Homespun Rat 'n' Toad Pie' and serves 4 people. Change these amounts so there's enough for SIX."

4 Freshly-caught Rats (frozen will do)
2 Common Toads (minus warts)
5 Ounces of 'Froggatt's Lumpy Sprout Ketchup'
8 Soiled Potatoes
A big wodge of pastry

Answer

DIVIDE each amount to get enough for one person, then TIMES to get enough for SIX. (So divide by 4 then times by 6.)

4 Rats ÷ 4 = 1 Rat (for one person) ×6 = 6 Rats (for 6 people)

2 Toads ÷ 4 = 0.5 Toad (for one person) ×6 = 3 Toads (for 6 people)

5 Ounces of 'Froggatt's Lumpy Sprout Ketchup' ÷ 4
= 1.25 Ounces (for one person) ×6 = 7.5 Ounces (for 6 people)

8 'taties ÷ 4 = 2 'taties (for one person) ×6 = 12 'taties (for 6 people)

A big wodge of pastry ÷ 4 then ×6 = A wodge of pastry half as big again.

In fact, all the amounts are just HALF AS MUCH AGAIN, if you notice.

The Three Mathsketeers say "divide for one, then times for all"...

Smug gits. It's a simple rule — the trick is knowing when to use it. Learning the examples above will help.

1) If seven pencils cost 98p, how much will 4 pencils cost?
2) Work out the amount of each ingredient needed to make enough Rat 'n' Toad pie for 9 people.*

*If you could e-mail me your answer, that would be great — I'm holding a dinner party next Saturday.

Ratios

A favourite type of question they like to ask you in Exams is comparing the 'value for money' of 2 or 3 similar items. Always follow the GOLDEN RULE...

Divide by the PRICE in pence (to get the amount per penny)

Example: The local 'Supplies 'n' Vittals' stocks three sizes of Jamaican Gooseberry Jam. The question is: Which of these represents 'THE BEST VALUE FOR MONEY'?

500 g at £1.08
350 g at 80p
100 g at 42p

ANSWER: DIVIDE BY THE PRICE IN PENCE TO GET THE AMOUNT PER PENNY

So we shall:

500 g ÷ 108p = 4.6 g PER PENNY
350 g ÷ 80p = 4.4 g PER PENNY
100 g ÷ 42p = 2.4 g PER PENNY

THE 500 g JAR is the best value for money because you get MORE JAM PER PENNY. With any question comparing 'value for money', DIVIDE BY THE PRICE (in pence) and it will always be the BIGGEST ANSWER is the BEST VALUE FOR MONEY.

The whole grisly subject of RATIOS gets a lot easier when you: **Turn RATIOS into FRACTIONS**

What the fraction form of the Ratio actually means

1) Suppose in a class there's girls and boys in the ratio 3 : 4. This means there's ¾ as many girls as boys.
2) So if there were 20 boys, there would be ¾ × 20 = 15 girls. You've got to be careful though — it doesn't mean ¾ of the people in the class are girls. In fact, three sevenths of the class are girls.

Reducing Ratios to their simplest form

You reduce ratios just like you'd reduce fractions to their simplest form. For the ratio 15:18, both numbers have a factor of 3, so divide them by 3 — that gives 5:6. We can't reduce this any further. So the simplest form of 15:18 is 5 : 6.

Treat them just like fractions — use your calculator if you can

Now this is really sneaky. If you stick in a fraction using the $a^b/_c$ button, your calculator automatically cancels it down when you press =. So for the ratio 8:12, just press 8 $a^b/_c$ 12 =, and you'll get the reduced fraction $\frac{2}{3}$. Now you just change it back to ratio form ie. 2 : 3. Ace.

The other golden rule — never dunk a biscuit for longer than 3 seconds...

Learn the Golden Rule for finding a best buy, then cover it up and answer these little beauties:

1) 'Slugtail Soup' comes in three different sizes: 150 g at 87p, 250 g at £1.37 and 750 g at £3.95. Work out which one is the best value for money. (And don't just guess!)
2) Reduce the ratio 21 : 56 to its simplest form.

Revision Test for Unit 1 — 1

WHAT YOU'RE SUPPOSED TO DO HERE is put all the methods of this Unit so far into practice to answer these questions.

1) These are the sort of numbers you might get in your calculator display:
a) 1.2343534 b) 2.9999999 c) 15.534624 d) 12.0833
Round them off to the nearest whole number.

2) Round these numbers off to 1 decimal place: a) 5.32 b) 3.46 c) 6.15

3) Round these numbers off to 2 decimal places: a) 5.327 b) 3.463 c) 6.159

4) Round these numbers off to 3 decimal places: a) 5.3272 b) 3.4638 c) 6.1594

5) a) Give 246 to the nearest 10 b) Give 860 to the nearest 100

6) How many significant figures have these numbers got?
a) 12 b) 150 c) 2000 d) 23.4 e) 8500

7) Use the $a^b/_c$ button to reduce $^{12}/_{15}$ to its simplest form.

8) Calculate a) 4/7 of 560 b) 2/5 of £150 c) 65% of 300

9) A DVD player costs £50 plus VAT. If VAT is 17.5%, how much does the DVD player cost?

10) Carl has £35 to spend. He wants to use a 20%-off voucher to buy a top that should cost £45. Can he afford the top?

11) What is The Golden Rule for ratios?

12) Last week Rick ordered 5 pints of milk from the milkman. His bill was £2.35. This week he orders 3 pints of milk. How much will this week's bill be?

13) What is The Golden Rule for finding the 'Best Buy'?

14) Charley loves ham. Two different sized tins of Froggatt's ham are on sale in his local shop. Which one is the 'Best Buy' for Charley?

15) Use your calculator to reduce these ratios to their simplest form:
a) 13 : 117 b) 95 : 152 c) 54 : 81

Collecting Data

All stats problems start with something to be tested called a hypothesis, e.g. "more people are driven to school than walk". Once you have your hypothesis, you can start to collect data to see if it's true...

Know your Data Types

The first thing to do when you're collecting data is to identify exactly what data you need.

1) PRIMARY data is data you've collected yourself (e.g. from a survey you've done).
2) SECONDARY data is collected by someone else (e.g. data you can find in newspapers or on websites).
3) QUALITATIVE data is data that uses words to describe it — not numbers. E.g. favourite pizza topping.
4) QUANTITATIVE data measures quantities and is anything you can measure with a number. E.g. heights.
5) Quantitative data is either discrete or continuous:

DISCRETE DATA is data that can be measured exactly.

E.g. The number of goals scored in a game.

CONTINUOUS DATA is data that can take any value in an interval.

E.g. The height of a teenager.

Questionnaires Should be Designed Carefully

Questionnaires are a common way of collecting primary data, which means they're a favourite with examiners. You might be asked to comment on how good a question is, or write some questions of your own. Either way, you need to bear these points in mind:

1. **MAKE SURE YOUR QUESTIONS ARE RELEVANT**

   It's no good asking really fascinating questions if the answers aren't going to be useful.

2. **QUESTIONS SHOULD BE CLEAR, BRIEF AND EASY TO UNDERSTAND**

   Your best bet is to assume that the people answering them are really stupid.

3. **ALLOW FOR ALL POSSIBLE ANSWERS TO YOUR QUESTION**

   E.g. "What is your favourite subject: Maths, English or Science?" is difficult to answer truthfully if you like Art best — to help, you could add an "other" category.

4. **QUESTIONS SHOULDN'T BE LEADING OR BIASED**

   Leading or biased questions are ones that suggest what answer is wanted.
   For example: "You do agree that thrash metal is really good music?"
   The problem with this question is that it could make the interviewee feel pressurised into saying 'yes'. A better question would be "What type of music do you prefer to listen to?"

5. **QUESTIONS SHOULD BE UNAMBIGUOUS**

   Here's an example: "Do you play computer games a lot?"
   This question could be interpreted differently by different people. One person could answer yes, while another who plays the same amount could answer no. A better question would be "How many hours do you play computer games per week?" because it isn't open to different interpretations.

6. **QUESTIONS SHOULDN'T BE VAGUE**

   If questions are vague you won't get the results you need to answer your question. Here's an example: Julie is carrying out a survey on what music students like to listen to, her first question is "Do you like music?". It's really vague and the answers people give won't be useful to her.
   A better question is "What type of music do you like to listen to?".

Collecting Data

Another page on collecting data — this time designing collection tables and answers. Ace...

Make Sure Observation Sheets have Enough Space

If you're carrying out an experiment, you'll need an observation sheet to record your results.
Make sure the sheet has enough space for all the data you want to record and is set out in a logical way:

EXAMPLE 1: Dave is carrying out a survey to record how much 20 students in his school spend on food and drink at lunchtime. He needs enough space to write down how much 20 people spend on both food and drink, so you'd need something like this but with 20 rows (instead of the 3 here).

Student	Food (£)	Drink (£)	Total (£)
1	2.52	0.95	3.47
2			
3			

EXAMPLE 2: You could use a simple tally table to collect data on coin tossing.

Heads	Tails
𝍸 I	IIII
6	4

Survey Questions can be Open or Closed

Closed questions have a fixed number of possible answers — these could be yes/no or tick box questions.

The good thing about closed questions is that you can easily process the data collected. Also, if the question is well designed, the responses won't be ambiguous at all. But, the answers are limited to the options given.

Keep an eye out for survey responses that don't offer enough choice or overlap — these are badly designed.

1) Are you under 18 years of age?
2) Tick the mode of transport you use to get to school.

Bus ☐ Car ☐ Bicycle ☐
Walking ☐ Other ☐

You can Split your Data into Classes

1) If you're collecting lots of data, or your data's spread out over a large range, you can make it more manageable by grouping it into different classes.
2) When you do this, it's important that you define the classes well so none of them overlap — this means that each bit of data can only be put into one class.

Age in completed years	0 – 19	20 – 39	40 – 59	60 – 79	80 – 99
Number of people	6	13	14	8	9

3) The problem with grouping data is that you lose some of the accuracy of the data because you don't know what the exact data values are any more.

Who wants to collect a questionnaire...

... is the (not so exciting) quiz spin-off. Make sure you learn these two pages, then try this question...

1) Give one criticism of each of these questions: a) Do you watch a lot of television?
   b) Do you agree that maths is the most important subject taught in schools?
   c) What is your favourite drink? Answer A, B or C. A) Tea B) Milk C) Coffee

Sampling

Sampling — Cheaper and Easier than Asking Everyone

1) For any statistical project, you need to find out information about a group of people or things. This group is called the POPULATION. Examples of populations are things like all the pupils in a school or all the people who have access to the internet.
2) You collect information about a population by doing a SURVEY — it can be used to collect data about the whole population or from a SAMPLE (part of the population).
3) Surveying a sample is a relatively easy way of finding out about a population. You can use the data you collect to make estimates and draw conclusions about the whole population.
4) There are pros and cons of using sample data:.

PROS	It's a lot quicker, cheaper and often more practical than doing a survey of the entire population.
CONS	You don't have information about every member of the population.

Make Sure your Data is Reliable and Not Biased

1) When you sample a population it's important to make sure the sample fairly represents the whole population. This means any conclusions you draw from the data in the sample can be applied to the whole population.
2) A BIASED study is one that doesn't fairly represent the whole population. To avoid bias you need to:
   - Use a REPRESENTATIVE SAMPLE — in other words, make sure that your sample includes people from the population you're interested in and that no-one is excluded.
   - Select your sample at RANDOM. Something is chosen at random when every item in a group has an equal chance of being chosen.
3) A bigger sample is better because it's more likely to be representative. It'll provide more reliable data — but it might be less practical to collect.

Spotting Problems With Sampling Methods

All sampling methods can be affected by bias. So, to make sure you can spot a biased sample a mile off, here's a lovely example...

EXAMPLE: In a telephone poll, 100 people were asked if they use the train regularly and 20% said yes. Does this mean 20% of the population regularly use the train?

ANSWER: Probably not. There are several things wrong with this sampling technique:

- First and worst: the sample is far too small. At least 1000 would be more like it.
- What about people who don't have their own phone?
- What time of day was it done? When might regular train users be in or out?
- Which part or parts of the country were telephoned?
- Were they sampled at random?

When getting a sample — size matters...

...so that it's more representative. Have a go at this question: explain whether a survey on the music listening preferences of a school that uses a Year 7 class as a sample is likely to be representative.

Mean, Median, Mode and Range

If you don't manage to learn the 4 basic definitions then you'll be passing up on some of the easiest marks in the whole Exam. It can't be that difficult can it?

When you've got grouped data, the mode is called the modal class (see p. 22)

1) MODE = MOST common

Mode = most (emphasise the 'o' in each when you say them)

2) MEDIAN = MIDDLE value

Median = mid (emphasise the m*d in each when you say them)

3) MEAN = TOTAL of items ÷ NUMBER of items

Mean is just the average, 'but it's mean 'cos you have to work it out'

4) RANGE = How far from the smallest to the biggest

THE GOLDEN RULE:

Mean, median, mode and range should be easy marks but even people who've gone to the incredible extent of learning them still manage to lose marks in the Exam because they don't do this one vital step:

Always REARRANGE the data in ASCENDING ORDER

(and then check you have the same number of entries)

Example:

"Find the mean, median, mode and range of these numbers:"

2, 5, 3, 2, 6, -4, 0, 9, -3, 1, 6, 3, -2, 3 (14)

1) FIRST... rearrange them: -4, -3, -2, 0, 1, 2, 2, 3, 3, 3, 5, 6, 6, 9 (✓14)

2) MEAN $= \frac{\text{total}}{\text{number}} = \frac{-4-3-2+0+1+2+2+3+3+3+5+6+6+9}{14}$

$= 31 \div 14 = 2.21$

3) MEDIAN = the middle value (only when they're arranged in order of size, that is).

When there are two middle numbers, as in this case, then the median is halfway between the two middle numbers.

-4, -3, -2, 0, 1, 2, 2, 3, 3, 3, 5, 6, 6, 9

← seven numbers this side   seven numbers this side →

Median = 2.5

4) MODE = most common value, which is simply 3. (Or you can say "The modal value is 3")

5) RANGE = distance from lowest to highest value,
i.e. from -4 up to 9, = 13

Don't forget the order — socks and then shoes...

Trust me — it's easily done. Learn the four definitions and the Golden Rule, then cover this page and write them all down from memory. Then be a clever bunny and use everything you've learnt to find the mean, median, mode and range for this set of data: 1, 3, 14, -5, 6, -12, 18, 7, 23, 10, -5, -14, 0, 25, 8

Averages

Average is just general term for mean, median and mode. And as a reward for learning that exciting fact, here's a page about using averages to compare data.

Using Averages to Make Comparisons

Example:

Some Z-list celebrities took part in a new TV show "Wuff Factor". The celebrities were split into two teams (A and B) and had to take their pet dogs through various tests.

Team A Marks
3
6
6
7
9
9
10
14

PART 1 — THE AGILITY TEST
The table shows the marks, out of 20, gained by the Team A dogs in the agility test.

Q1: Use the information above to calculate the mean and median marks for Team A.

ANSWER: TEAM A MEAN = total of marks ÷ number of dogs = 64 ÷ 8 = 8
MEDIAN is the middle mark. There are 8 marks, so the middle is halfway between the 4th and 5th marks:

3, 6, 6, (7, 9), 9, 10, 14 = (7 + 9) ÷ 2 = 8

3 numbers on each side.

TEAM B'S RESULTS: Team B got the following results **Mean = 15, Median = 16**

Q2: Which team did better? Explain your answer.

ANSWER: So you just need to compare the mean and median values for both teams. Your answer should go something like this...
'Team B did better because their mean and median marks were both much higher.'

Using Range to Make Comparisons

Example:

12.5	12.8	13.4	14.0	14.2	14.2	15.6	16.5

PART 2 — THE BEHAVIOUR TEST
The scoreboard shows the scores gained by the Team A dogs in the behaviour test.

Q1: Calculate the range of scores for Team A.

ANSWER: RANGE = biggest – smallest = 16.5 – 12.5 = 4.0

Q2: The range of scores for Team B was 9.2.
Say, with a reason, which team was more consistent?

ANSWER: To answer this you just need to compare the ranges. Makes sense if you think about it. The range for Team A is 4.0, which means that all the scores lie within 4 marks of each other — pretty consistent. For Team B, the range of scores is 9.2 marks. So...
'Team A was more consistent because their range of scores was much smaller.'

There's nothing average about CGP books...

Have a good read of the examples here, then check out this teaser.

Variety X: Mean = 15, range = 4
Variety Y: Mean = 10, range = 7

1) Geoff has two different varieties of potato plant — X and Y. The mean number of potatoes per plant, and the range for each variety are shown above. Geoff's friend Sid asks Geoff which variety he would recommend. Say, with reasons, which variety Geoff should choose.

Frequency Tables

Frequency tables can either be done in rows or in columns of numbers. They can be quite confusing, but they're not too bad if you learn these key points:

1) The word FREQUENCY just means HOW MANY, so a frequency table is nothing more than a 'How many in each group' table.
2) The FIRST ROW (or column) just gives the GROUP LABELS.
3) The SECOND ROW (or column) gives the ACTUAL DATA.
4) To find the MEAN you have to WORK OUT A THIRD ROW (or column) and then **MEAN = 3rd Row total ÷ 2nd Row Total**

Example:

Here is a typical frequency table in ROW FORM and COLUMN FORM:

Row Form

No. of Sisters	0	1	2	3	4	5	6
Frequency	7	15	12	8	3	1	0

Column Form

No. of Sisters	Frequency
0	7
1	15
2	12
3	8
4	3
5	1
6	0
Totals	46

The mode and range are easy to find from the table:

The MODE is just the group with the most entries: i.e 1

The 2nd row tells us there are people with anything from 'no sisters' right up to 'five sisters' (but not 6 sisters). So the RANGE is 5 – 0 = 5

To find the mean for the data, you need to add an extra row:

No. of Sisters	0	1	2	3	4	5	6	Totals	
Frequency	7	15	12	8	3	1	0	46	(People asked)
No. × Frequency	0	15	24	24	12	5	0	80	(Sisters)

Now from the table:

MEAN $= \dfrac{\text{3rd row total}}{\text{2nd row total}} = \dfrac{80}{46} =$ 1.74 (sisters per person)

MEDIAN — imagine the original data SET OUT IN ASCENDING ORDER:

0000000 111111111111111 222222222222 33333333 444 5

The median is just the middle number — which is between the 23rd and 24th digits. So it's 2.

Mean sisters

My table has 5 columns, 6 rows and 4 legs...

Learn the rules for frequency tables, then turn over and write them down to see what you know. The best way of getting your head round this is to practise doing it, so have a go yourself. You could ask your class how many times they've been to Bury. Then put the results in a frequency table.

Grouped Frequency Tables

Frequency tables often group data together to make it easier to understand. I, on the other hand, like to write with my eyes closed to make it harder to understand. But that's just me.

Example 1:

The marks of 28 students in a test (out of 80) were:

63, 45, 44, 52, 58,
49, 48, 22, 37, 34,
44, 49, 66, 73, 69,
32, 49, 29, 55, 57,
30, 72, 59, 46, 70,
39, 27, 40

As a Grouped Table

Marks	Tally	Frequency
$0 \le x \le 10$		
$11 \le x \le 20$		
$21 \le x \le 30$	IIII	4
$31 \le x \le 40$	~~IIII~~	5
$41 \le x \le 50$	~~IIII~~ III	8
$51 \le x \le 60$	~~IIII~~	5
$61 \le x \le 70$	IIII	4
$71 \le x \le 80$	II	2
		28

Example 2:

The weights (in kg) of a bunch of 20 school kids are shown below.

67.3, 45.6, 47.7, 65.0, 54.2,
76.5, 44.6, 34.3, 69.8, 53.9,
32.3, 54.5, 78.9, 59.8, 57.4,
30.0, 79.1, 46.2, 66.0, 51.6

As a Grouped Table

Weight w (kg)	Tally	Frequency
$30 \le w < 40$	III	3
$40 \le w < 50$	IIII	4
$50 \le w < 60$	~~IIII~~ I	6
$60 \le w < 70$	IIII	4
$70 \le w < 80$	III	3
		20

Reading the Intervals

In the top table, '$0 \le x \le 10$' means x is either between 0 and 10 or it is one of those values.

In the bottom table, '$30 \le w < 40$' means w is between 30 and 40 or it could equal 30, but it can't equal 40 (40 would go in the next group).

The intervals used in the top table are suitable for whole number data (note that you couldn't place 30.5 on this table — it would fall in between two groups). Whole numbers are an example of discrete data — data that can only take certain values.

The intervals in the bottom table are suitable for any numbers (30.5, 50.9999) because there are no gaps between the groups. Data like this that can take any values within a range is called continuous data.

Discretion is the better part of valour...

Learn the difference between discrete and continuous data — you'll lose marks if you get them mixed up. Think of your own examples for each type — e.g. the number of celebrities I stalk each day is discrete data, and the length of time it takes the police to arrive to arrest me is continuous. On a slightly less creepy note... 1) Say if the following are continuous or discrete data and design a data table for each one: a) shoe size of 20 adults b) height of 30 adults.

Grouped Frequency Tables

You'll be pleased to know that this is the last page to endure on frequency tables, woohoo... But there are some other types of table just around the corner. I knew it was too good to be true.

Estimating The Mean For Grouped Data

You need to be able to estimate the mean for data in a grouped frequency table. Note you can only estimate the mean because you don't know the actual values.

The method is a bit fiddly at first, but it's easy once you've learnt it:

1) Add a 3rd row and enter MID-INTERVAL VALUES for each group.
2) Add a 4th row and multiply FREQUENCY × MID-INTERVAL VALUE for each group.
3) Work out the TOTALS of rows 2 and 4.
4) Get the mean by dividing ROW 4 TOTAL by ROW 2 TOTAL.

To find the mid-interval value, you just add together the upper and lower limits and divide by 2.

Note — if your table is arranged like the ones on the previous page, you'll need to add extra columns rather than rows...

Example:

The table below shows the distribution of weights of 60 children. Find the modal class, median and estimate the mean.

Weight	$30 \le w < 40$	$40 \le w < 50$	$50 \le w < 60$	$60 \le w < 70$	$70 \le w < 80$
Frequency	8	16	18	12	6

1) The modal class is just the group with the highest frequency: $50 \le w < 60$ kg
2) The median can't be found exactly but you can say which group it's in. If all the data were put in order, the weights in the middle (30th & 31st entries) would be in the $50 \le w < 60$ kg group.
3) To find the mean, add two rows to the table as described above:

Weight (kg)	$30 \le w < 40$	$40 \le w < 50$	$50 \le w < 60$	$60 \le w < 70$	$70 \le w < 80$	Totals
Frequency	8	16	18	12	6	60
Mid-Interval Value	35	45	55	65	75	—
Frequency × Mid-Interval Value	280	720	990	780	450	3220

4) Now, just divide the totals to get an estimate of the mean:

$$\text{Mean} = \frac{\text{Overall Total (Final Row)}}{\text{Frequency Total (2nd Row)}} = \frac{3220}{60} = 53.7$$

This page auditioned for Britain's next top modal...

It didn't get through — but I think I'll try again next year. Learn all the stuff on this page, then turn over and write down everything you've just learned. Good, clean fun. Then have a go at these...
1) Estimate the mean for this table: 2) State the modal class. 3) Which group contains the median?

Length L (cm)	$15.5 \le L < 16.5$	$16.5 \le L < 17.5$	$17.5 \le L < 18.5$	$18.5 \le L < 19.5$
Frequency	12	18	23	8

Tables, Charts and Graphs

Make sure you know how to represent your data in different types of charts, tables and graphs.

1) Line Graphs and Frequency Polygons

A line graph is just a set of points joined with straight lines.

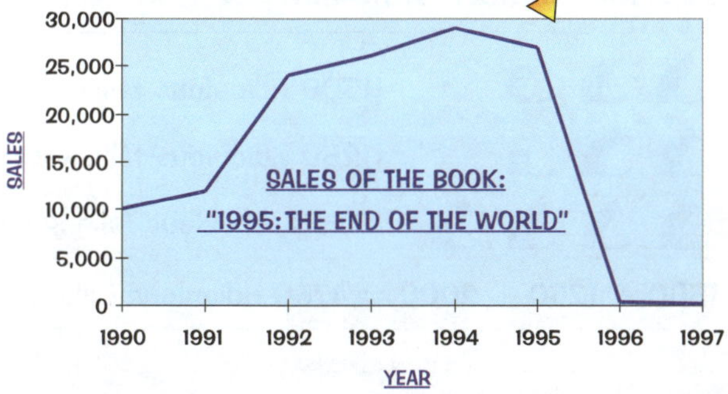

A frequency polygon looks similar and is used to show the information from a frequency table like the ones on p.20.

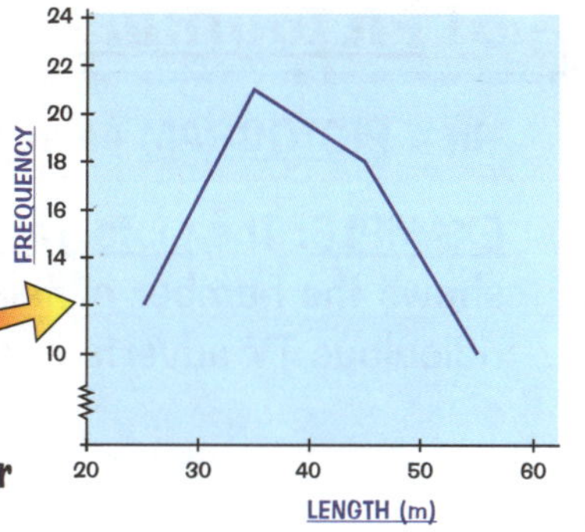

Frequency polygons for grouped data use the midpoint of each class.

EXAMPLE This frequency table shows the heights of 12 plants in a biology experiment. Draw a frequency polygon using this data.

Height	$0 \le h < 5$	$5 \le h < 10$	$10 \le h < 15$	$15 \le h < 20$
Frequency	3	5	2	2

1) You need to work out the MIDPOINTS for each class of heights first:

Height	$0 \le h < 5$	$5 \le h < 10$	$10 \le h < 15$	$15 \le h < 20$
Midpoint	2.5	7.5	12.5	17.5
Frequency	3	5	2	2

2) Then PLOT the frequency of each class against its midpoint and join the points with straight lines (because it's a polygon rather than a curve).

Frequency is always on the y-axis.

The highest peak tells you the group with the highest frequency — 5-10 cm class.

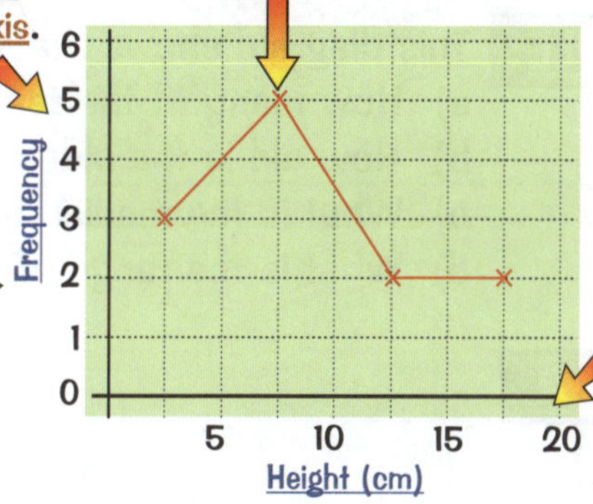

Values on the x-axis are the class boundaries.

2) Two-Way Tables

These are a bit like frequency tables but they show two different things:

	Women	Men	TOTAL
Left-handed		27	63
Right-handed	164	173	
TOTAL	200	200	400

EXAMPLE: "Use this table to work out how many (a) right-handed people and (b) left-handed women there were in this survey."

ANSWER:

(a) Either: (i) add up the number of right-handed women and the number of right-handed men. So that's 164 + 173 = 337 right-handed people.
Or: (ii) take away the total number of left-handed people from the total number of people. So that's 400 – 63 = 337 right-handed people.

	Women	Men	TOTAL
Left-handed		27	63
Right-handed	164	173	
TOTAL	200	200	400

(b) Either: (i) take away the number of right-handed women from the total number of women. That's 200 – 164 = 36 left-handed women.
Or: (ii) take away the left-handed men from the total number of left-handed people. Which would be 63 – 27 = 36 left-handed women.

	Women	Men	TOTAL
Left-handed		27	63
Right-handed	164	173	
TOTAL	200	200	400

No animals were harmed in the making of the lion graph above...

Ho. Ho. Have a go at sketching examples of each type of table or graph — just make up some data. Then copy out the two-way table, close the book and fill in the blanks. Check the numbers add up.

Tables, Charts and Graphs

Ooo, now it's time for a graph where you get to draw pretty pictures. And some other types, but they're not as much fun. You've still got to know them though.

3) Pictograms — these use pictures instead of numbers.

In a PICTOGRAM each picture or symbol represents a certain number of items.

EXAMPLE: The pictogram opposite shows the number of talking cats used in ridiculous TV adverts in a 3-month period:

= 500 talking cats

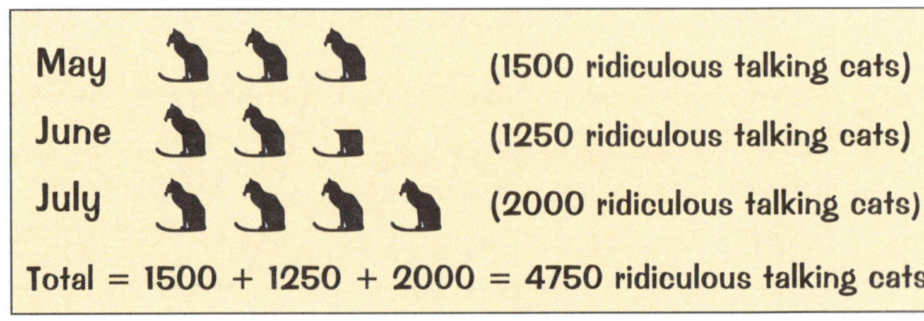

4) Stem and Leaf Diagrams

Stem and leaf diagrams are a bit like bar charts (see p.26), but more confusing. They're supposed to be easy to read, but they're not. So LEARN this example.

EXAMPLE: This diagram shows the ages of my school teachers.

a) How many of the teachers are in their forties?
b) How old is the oldest teacher?
c) What is the median age?
d) What is the range of ages?

Stem	Leaf
3	3 5
4	0 5 7 8
5	1 4 9
6	1 3

Key: 5 | 4 means 54

The key tells you how to read the diagram. A 5 in the stem and a 4 in the leaf means 54.

ANSWER:

Step 1:
Write down all the ages of the teachers, using the key.

33, 35,
40, 45, 47, 48,
51, 54, 59,
61, 63

Step 2:
Answer the question.

a) four
b) 63
c) 48
d) 30

Range is simple — you can easily see the smallest and largest ages for each class.

So, the range is = 63 – 33 = 30 years.

The median age of the teachers is the middle value (or halfway between the two middle values) — it's easy to read these values off the diagram as the ages are already presented in ascending order.

If stem and leaf diagrams be the food of love, I wanna be single...

Ey up, that was a bit cultured wasn't it. Learn all about pictograms and stem and leaf diagrams, then...

1) Draw a stem and leaf diagram for this data: 3, 16, 14, 22, 7, 11, 26, 17, 12, 19, 20, 6, 13, 24, 26
2) Find the median and range of the data set.

Tables, Charts and Graphs

You use a Scatter graph to show how two different types of data (like height and weight) are related to each other. And as if by magic, there are some examples below.

5) Scatter Graphs

1) A SCATTER GRAPH is just a load of points on a graph that end up in a bit of a mess rather than in a nice line or curve.
2) There is a posh word to say how much of a mess they are in — it's CORRELATION.
3) Good Correlation (or Strong Correlation) means the points form quite a nice line, and it means the two things are closely related to each other. When this is the case, you can draw a line of best fit roughly through the middle of the scatter of points.

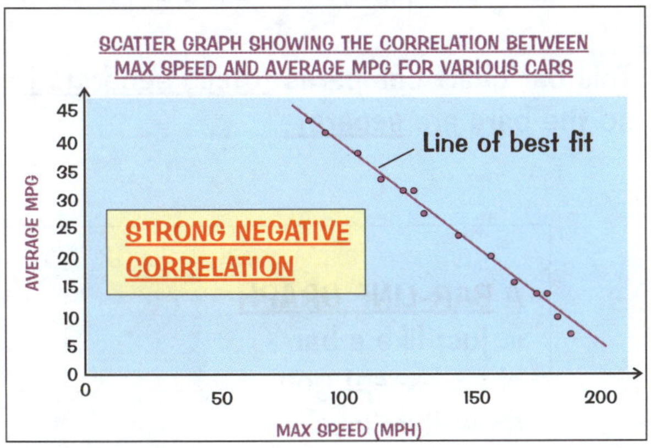

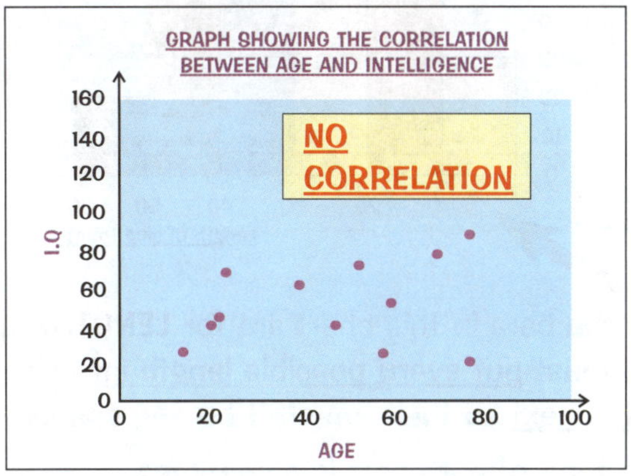

4) No Correlation means the points are all over the place and so there is no relation between the two things.
5) If the points form a line sloping UPHILL from left to right, then there is POSITIVE CORRELATION, which just means that both things increase or decrease together.
6) If the points form a line sloping DOWNHILL from left to right, then there is NEGATIVE CORRELATION, which just means that as one thing increases the other decreases.
7) So when you're describing a scatter graph you have to mention both things, i.e. whether it's strong/weak/moderate correlation and whether it's positive/negative.

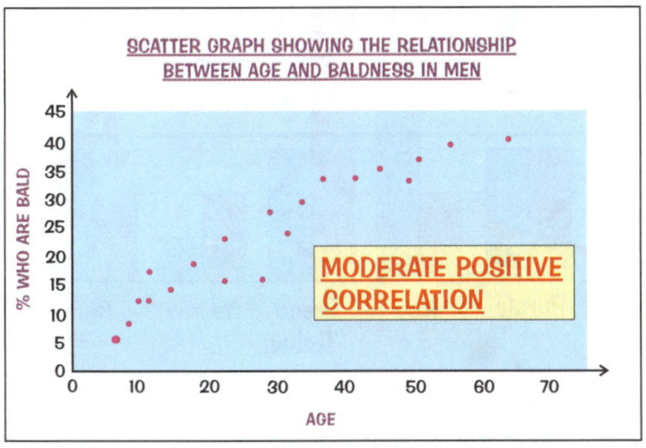

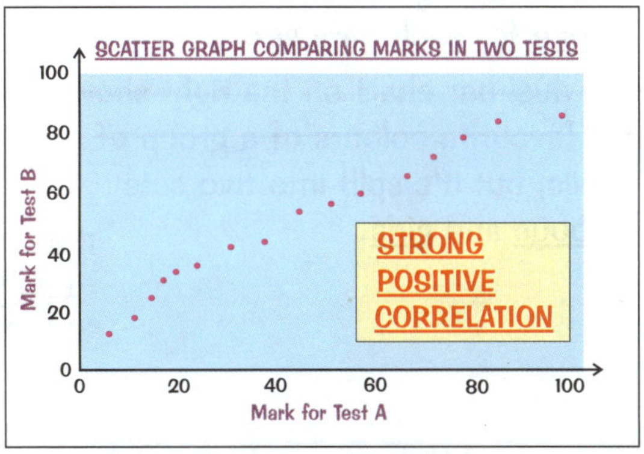

What soap do maths teachers watch — Correlation Street...

Recent studies have shown that there is strong positive correlation between time spent practising your evil laugh / cat stroking / beard combing and how successful you are as an evil genius. Learn all the graph on this page and have a go at drawing an example of each.

Bar Charts and Comparing Data

More charts and graphs... Oh yes, yes, yes, yes, yes...

Bar Charts

Just watch out for when the bars should touch or not touch:

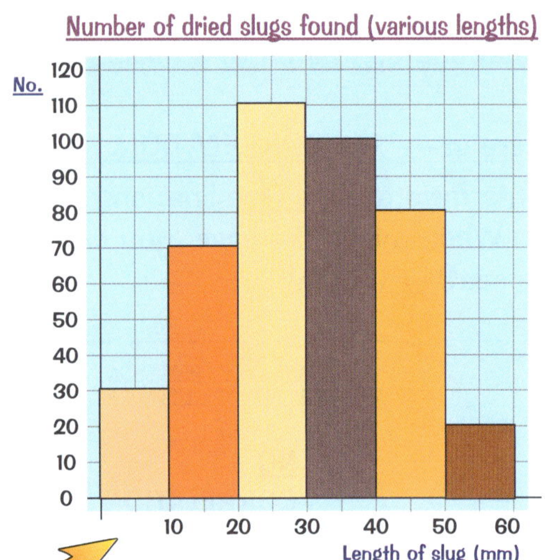

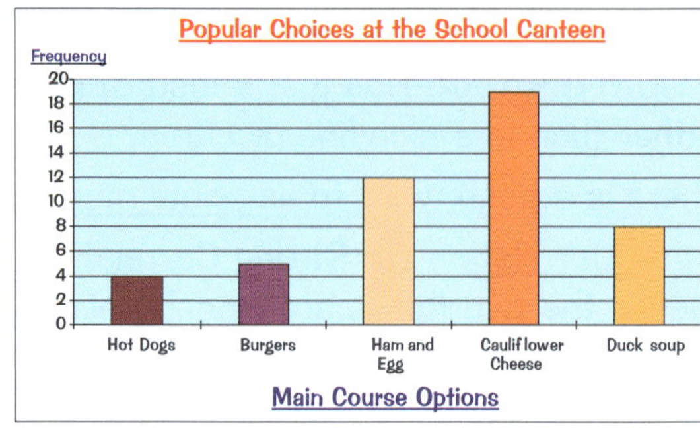

This bar chart compares totally separate items so the bars are separate.

ALL the bars in this chart are for **LENGTHS** and you must put every possible length into one bar or the next so there mustn't be any spaces.

This type of bar chart is sometimes called a histogram.

A **BAR-LINE GRAPH** is just like a bar chart except you draw thin lines instead of bars.

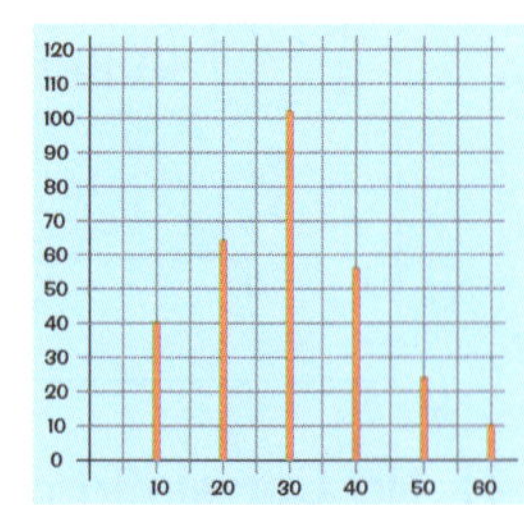

Dual Bar Charts Can be Used to Compare Data Sets

1) Dual bar charts show two sets of data at once so it's easy to compare them.
2) Each category has two bars — one for each data set.
3) The dual bar chart on the right shows the favourite colours of a group of pupils, but it's split into two sets — boys and girls.

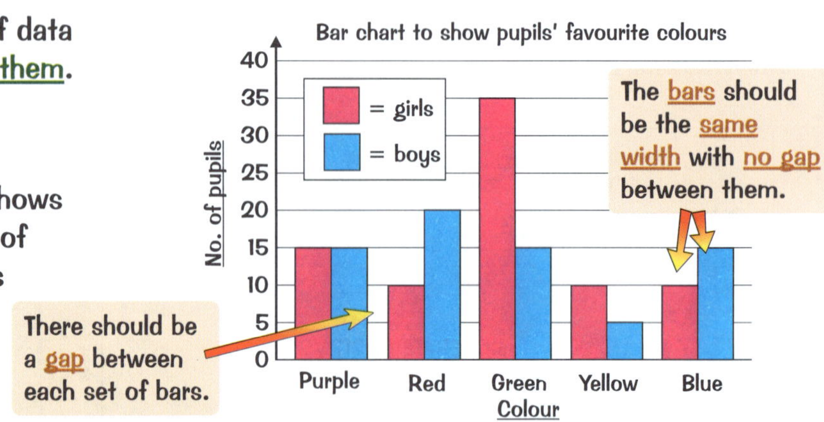

A sheep's favourite graph — the baaaa chart...

Woah, have another gander over this page, then test yourself with this delightful question.

1) Use the results from the "boring team names" quiz to draw a dual bar chart showing how the teams performed in each round.

Team	Green	Blue	Red	Yellow
Points — 1st round	24	15	20	25
Points — 2nd round	16	15	22	20

Pie Charts

In a similar way that you can make the correct combination of meat and pastry into a delicious pie, examiners can make Pie Charts into tricky exam questions. So learn the Golden Rule for Pie Charts:

The TOTAL of Everything = 360°

Remember that 360° is the trick for dealing with most Pie Charts

1) Relating Angles to Fractions — you should just know these ones straight off:

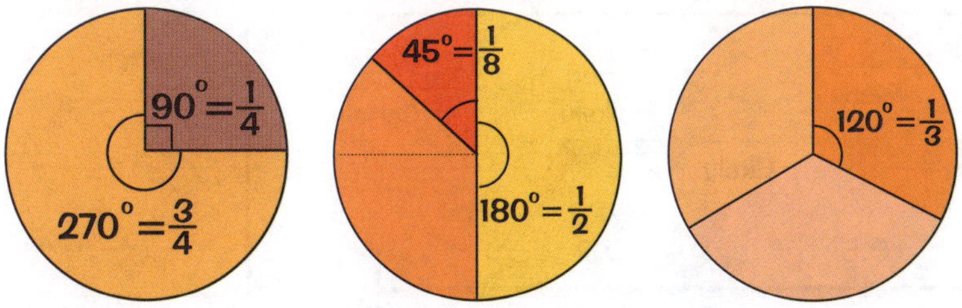

$90° = \frac{1}{4}$
$270° = \frac{3}{4}$

$45° = \frac{1}{8}$
$180° = \frac{1}{2}$

$120° = \frac{1}{3}$

For any angle the formula is:

$$\text{Fraction} = \frac{\text{Angle}}{360°}$$

And then cancel it down with your calculator (see p.6)

If you have to measure an angle, you should expect it to be a nice round number like 90° or 180° or 120°, so don't go writing 89° or 181° or anything silly like that.

2) Relating Angles to Numbers of Other Things

Creature	Stick insects	Hamsters	Guinea pigs	Rabbits	Ducks	Total
Number	12	20	17	15	26	90
	↓	↓ ×4	↓	↓	↓	↓ ×4
Angle		80°				360°

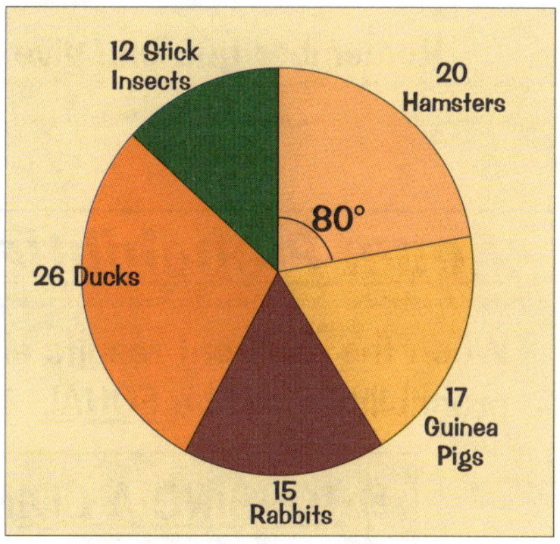

1) Add up all the numbers in each sector to get the TOTAL (=90).
2) Then find the MULTIPLIER (or divider) that you need to turn your total into 360°. For 90 → 360 as above, the MULTIPLIER is 4.
3) Now MULTIPLY EVERY NUMBER BY 4 to get the angle for each sector. E.g. the angle for hamsters will be 20 × 4 = 80°.

3) Use the Total to Find the Other Numbers

1) In the pie chart on the right, the total number of pets = 9
2) So you simply divide 360° by 9 = 40°
3) Then you divide each of the given angles by 40°
4) The final bit of pie is what's left over: 9 – 2 – 3 = 4

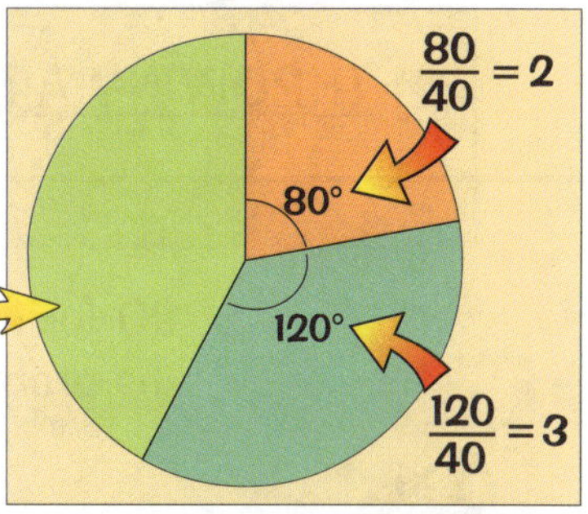

I bet you were expecting a joke about pies here...

... I'm far too witty to lower myself to making jokes about pies on a page about pie charts — that's much too obvious. There's just no need to go on and on about tasty steak-filled pies with scrumptious rich gravy on a bed of creamy mashed potato and a side order of mushy peas... mmm.

Anyway, while I'm off eating pies, you can display this data in a pie chart:

Football Team	Wigan A.	Luton	Man Utd	Others
No. of fans	53	15	30	22

Probability

Probability definitely seems a bit of a 'Black Art' to most people. It's not as bad as you think, but YOU MUST LEARN THE BASIC FACTS, which is what we have on these 3 pages.

All Probabilities are between 0 and 1

Probabilities can only have values from 0 to 1 (including those values). You should be able to put the probability of any event happening on this scale of 0 to 1.

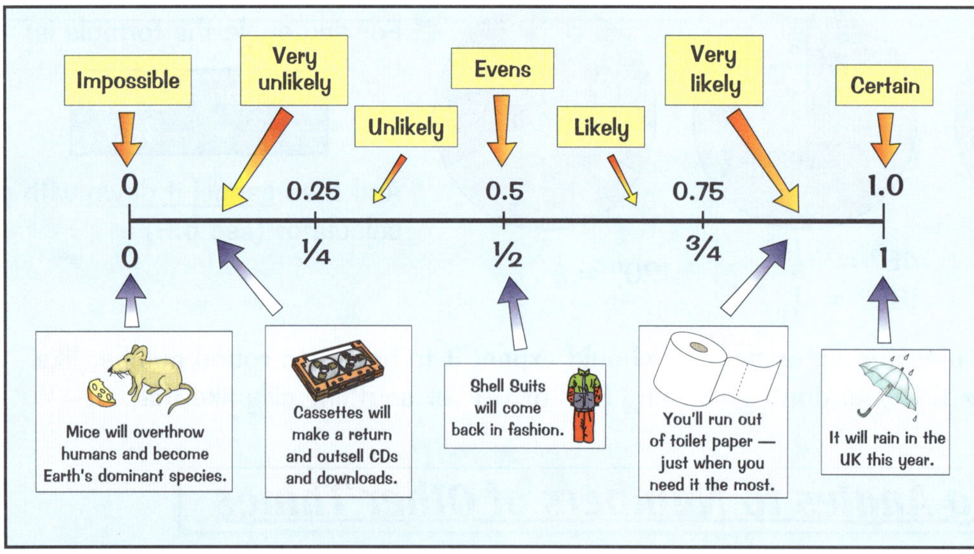

Remember you can give probabilities using FRACTIONS, DECIMALS or PERCENTAGES.

Equal Probabilities

When the different results all have the same chance of happening, then the probabilities will be EQUAL. These are the two cases which usually come up in Exams:

1) **TOSSING A COIN:** Equal chance of getting a head or a tail (probability $= \frac{1}{2}$)

2) **THROWING A DICE:** Equal chance of getting any of the numbers (probability $= \frac{1}{6}$)

SO for any event where the outcomes are equally likely, the probability of one outcome can be found by:

$$\text{Probability of one outcome} = \frac{1}{n}$$

(where n is the number of possible outcomes)

The probability of death and taxes is 1...

Picking cards from a pack is another example of equal probabilities, and it might come up in your Exam.

1) What is the probability of picking from a shuffled deck of cards (no jokers)
   a) An Ace of any suit? b) A number less than 7? c) a red picture card? (remember, it's 52 cards in a pack).

Probability

There's more to probability than just tossing coins and rolling dice you know. Oh yes.

Unequal Probabilities

These make for more interesting questions. (Which means you'll get them in the Exam.)

EXAMPLE 1: "A bag contains 6 blue balls, 5 red balls and 9 green balls. Find the probability of picking out a green ball."

ANSWER: The chances of picking out the three colours are NOT EQUAL.
The probability of picking a green is simply:

$$\frac{\text{NUMBER OF GREENS}}{\text{TOTAL NUMBER OF BALLS}} = \frac{9}{20}$$

EXAMPLE 2: "What is the probability of winning £45 on this spinner?"

ANSWER: The pointer has the same chance of stopping on every sector... ... and since there are 2 out of 8 which are £45 then it's a 2 out of 8 chance of getting £45. BUT REMEMBER ... you have to say this as a FRACTION or a DECIMAL or a PERCENTAGE:

2 out of 8 is 2 ÷ 8 which is 0.25 (decimal) or $\frac{1}{4}$ (fraction) or 25% (percentage)

The Probability of the Opposite happening is just the rest of the probability that's left over

If the probability of something happening is, say, 0.3 then the chance of it NOT HAPPENING is 1 – 0.3 (= 0.7), i.e. it's what's left when you subtract it from 1.

Example: A loaded dice has a 0.25 chance of coming up TWO.
What is the chance of it not coming up TWO?

Answer: 1 – 0.25 = 0.75. So, the chance of the dice not coming up TWO is 0.75.

Listing All Outcomes: 2 Coins, Dice, Spinners

A simple question you might get is to list all the possible results from tossing two coins or two spinners or a dice and a spinner, etc. Whatever it is, it'll be very similar to these, so LEARN THEM:

The possible outcomes from TOSSING TWO COINS are:

HH HT TH TT

From TWO SPINNERS with 3 sides:

BLUE, 1	RED, 1	GREEN, 1
BLUE, 2	RED, 2	GREEN, 2
BLUE, 3	RED, 3	GREEN, 3

Try and list the possible outcomes METHODICALLY — to make sure you get them ALL.

You can use your list to find probabilities, e.g. there are four equally likely outcomes from tossing two coins, so the probability of each outcome is ¼ or 0.25 or 25%.

Spinners fashion tip #56 — gold hot pants are an absolute must......

Spinners and bags of balls come up a lot in the Exam. Don't panic if they use another example — you work it out in exactly the same way. Try this one: 1) What is the probability of picking a white puppy from a bag containing 3 black puppies, 4 brown puppies, 2 white puppies and one purple puppy?

Relative Frequency

This isn't the number of times your granny comes to visit. It's a way of working out probabilities. Since you asked, my granny visits twice a year. She says she'd like to visit more, but sleeping on the blow-up bed plays mischief with her bad back.

Fair or Biased?

The probability of rolling a five on a dice is $\frac{1}{6}$ — you know that each of the 6 numbers on a dice is equally likely to be rolled, and there's only 1 five.

BUT this only works if it's a fair dice. If the dice is a bit wonky (the technical term is 'biased') then each number won't have an equal chance of being rolled. That's where Relative Frequency comes in — you can use it to work out probabilities when things might be wonky.

Do the Experiment Again and Again and Again and Again

You need to do an experiment over and over again and then do a quick calculation. (Remember, an experiment could just mean rolling a dice.) Usually the results of these experiments will be written in a table.

You might also hear experiments called trials.

The Formula for Relative Frequency

$$\text{Probability of something happening} = \frac{\text{Number of times it has happened}}{\text{Number of times you tried}}$$

You can work out the relative frequency as a fraction but usually decimals are best.

You usually get slightly different results each time you repeat an experiment. The important thing to remember is:

The more times you do the experiment, the more accurate the probability will be.

Example:

Number of Times the dice was rolled	10	20	50	100
Number of threes rolled	2	5	11	23
Relative Frequency	$\frac{2}{10} = 0.2$	$\frac{5}{20} = 0.25$	$\frac{11}{50} = 0.22$	$\frac{23}{100} = 0.23$

So, what's the probability? We've got 4 possible answers, but the best is the one worked out using the highest number of dice rolls. This makes the probability of rolling a three on this dice 0.23.

And since for a fair, unbiased dice, the probability of rolling a three is $\frac{1}{6}$ (about 0.17), then our dice is probably biased.

Dice rolls — a crunchy pack lunch alternative...

Blast those wonky dice. That's what started all this. Still it's all bound to come up on the Exam. So...

1) A 3-sided spinner is spun 100 times – it lands on red 43 times, blue 24 times and green the other times. Calculate the relative frequency of each outcome.

Revision Test for Unit 1 — 2

WHAT YOU'RE SUPPOSED TO DO HERE is put all the methods of the second half of this Unit into practice to answer these questions.

1) List three things you need to bear in mind when designing a question for a questionnaire.

2) For this set of numbers:
2, 6, 7, 12, 3, 7, 4, 15
a) Find the MODE
b) Find the MEDIAN
c) Find the MEAN
d) Find the RANGE

3) In a frequency table what does $50 \le w < 60$ mean? Would you put 50 in this group? What about 60, would it go in this group or the next one up, $60 \le w < 70$?

4) Calum is writing an article on the Skelly Crag half-marathon for the local paper. He wants to include the average time taken. Use the times of all 1000 runners, recorded in the table below, to work out the mean time taken.

Time (min)	$60 < t \le 90$	$90 < t \le 120$	$120 < t \le 150$	$150 < t \le 180$	$180 < t \le 210$	$210 < t \le 240$
Frequency	15	60	351	285	206	83

5) What is this diagram called? How many angry customers were there on Thursday?

6) A newspaper has claimed that a study shows a strong positive correlation between eating cheese and having nightmares. The results of the study are shown on the right.
a) What does strong positive correlation mean?
b) Do you agree with the newspaper claim?

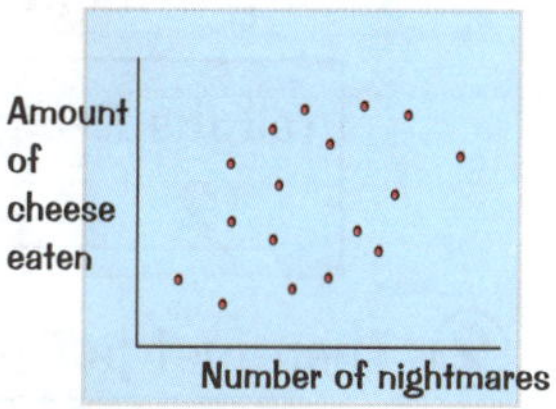

7) Fiona has carried out a traffic survey as part of her Geography course work. She wants to present the information in a pie chart.
a) Complete Fiona's table.
b) Draw the PIE CHART.

Colour	Blue	Red	Yellow	White	Totals:
Number of Cars	12	15	4	9	40
Angle on Pie Chart					360°

8) If I toss 2 coins, list all the possible outcomes. Since all these outcomes are equally likely, what is the chance of getting two heads?

9) If I toss a coin and throw a dice, list all the possible outcomes and say what the probability is of me getting a HEAD and a SIX.

10) The probability of a biased dice giving a SIX is 0.2. What is the chance of it NOT giving a six?

11) Colin is designing a game for the school summer fair. He wants the players to have a 50% chance of winning. He has a bag containing 3 red balls, 5 green balls and 7 black balls. If you pick out a red or green ball you win a prize.
a) What is the % chance of winning Colin's game?
b) How can he change the game so that there's a 50% chance of winning?

Big Numbers

Here's a nice easy page to get you going. You need to be able to:

1) Read big numbers E.g. how would you say 1 734 564?
2) Write them down E.g. write 'Thirty-two thousand and three' as a number.

Always Look at Big Numbers in Groups of Three

1) Always start from the extreme right-hand side of the number →
2) Moving left, ←, put a space in every 3 digits to break it up into groups of 3.
3) Now going right, →, read each group of three as a separate number and add 'million' and 'thousand' on for the first two groups (assuming 3 groups in all).

2 351 243

So many MILLION — So many THOUSAND — And the rest

(i.e. 2 million, 351 thousand, 243) or written fully in words:
Two million, three hundred and fifty-one thousand, two hundred and forty-three.

Putting Numbers in Order of Size

Example: 12 84 623 32 486 4563 75 2143

① It may not be exactly difficult, but it's still best to do it in two steps. First put them into groups, the ones with fewest digits first:

(all the 2-digit ones,	then all the 3-digit ones,	then all the 4-digit ones etc.)
12 84 32 75	623 486	4563 2143

② Then just put each separate group in order of size:

12 32 75 84	486 623	2143 4563

For decimals, do the whole-number bit first before looking at what's after the point. With numbers between 0 and 1, first group them by the number of 0s at the start. The group with the most 0s at the start comes first, just like this:

(those with 2 initial 0s,	then those with 1 initial 0,	then those with no initial 0s.)
0.0026 0.007	0.03 0.098	0.14 0.531 0.7

Once they're in groups, just order them by comparing the first non-zero digits.
(If the first digits are the same, look at the next digit along instead.)

Hint: Don't call numbers big or small to their face — they're very sensitive...

There's nothing too tricky about putting numbers into order of size — just remember the tips above. In fact, you might even find it strangely satisfying. A bit like alphabetising your CD collection.

1) Write these numbers in words: a) 1 234 531 b) 23 456 c) 2415 d) 3402 e) 203 412
2) Write this down as a number: Fifty-six thousand, four hundred and twenty-one
3) Put these numbers in order of size: 23 493 87 1029 3004 345 9
4) Write these numbers in ascending order: 0.37 0.008 0.307 0.1 0.09 0.2

Multiplying by 10, 100, etc

You really should know the stuff on this page because:
a) it's nice 'n simple, and b) they're likely to test you on it in the Exam.

1) To Multiply Any Number by 10

Move the Decimal Point ONE place BIGGER and if it's needed, ADD A ZERO on the end.

Examples:
23.6 × 10 = 236
345 × 10 = 3450
45.678 × 10 = 456.78

2) To Multiply Any Number by 100

Move the Decimal Point TWO places BIGGER and ADD ZEROS if necessary.

Examples:
296.5 × 100 = 29650
34 × 100 = 3400
2.543 × 100 = 254.3

3) To Multiply by 1000 or 10 000, the same rule applies:

Move the Decimal Point so many places BIGGER and ADD ZEROS if necessary.

Examples:
341 × 1000 = 341000
2.3542 × 10 000 = 23542

You always move the DECIMAL POINT this much:
1 place for 10, 2 places for 100,
3 places for 1000, 4 for 10 000 etc.

4) To Multiply by Numbers like 20, 300, 8000 etc.

MULTIPLY by 2 or 3 or 8 etc. FIRST, then move the Decimal Point so many places BIGGER (↷) according to how many noughts there are.

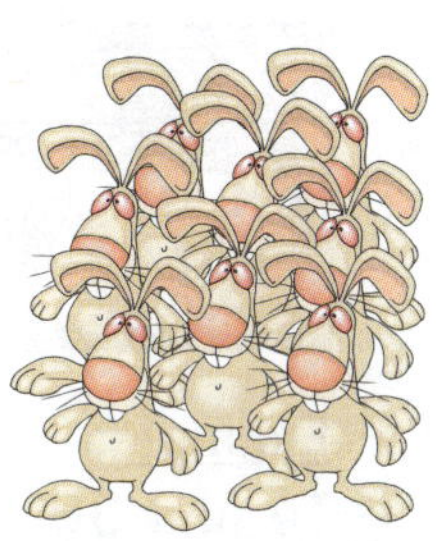

Example:
To find 234 × 200, first multiply by 2   234 × 2 = 468,
then move the DP 2 places   = 46800

Adding zeros when they're not needed? Tut, tut, noughty, noughty...

Just four multiplying methods to learn here — nothing too strenuous. For a bit of a workout, try these:

1) Work out a) 12.3 × 100 b) 345 × 10 c) 9.65 × 1000
2) Work out a) 2.4 × 20 b) 1.5 × 300 c) 60 × 3000

Dividing by 10, 100, etc

This is pretty easy stuff too. Just make sure you know it — that's all.

1) To Divide Any Number by 10

Move the Decimal Point ONE place SMALLER and if it's needed, REMOVE ZEROS after the decimal point.

Examples:

23.6 ÷ 10 = 2.36

340 ÷ 10 = 34

45.678 ÷ 10 = 4.5678

2) To Divide Any Number by 100

Move the Decimal Point TWO places SMALLER and REMOVE ZEROS after the decimal point.

Examples:

296.5 ÷ 100 = 2.965

340 ÷ 100 = 3.4

2543 ÷ 100 = 25.43

3) To Divide by 1000 or 10 000, the same rule applies:

Move the Decimal Point so many places SMALLER and REMOVE ZEROS after the decimal point.

Examples:

341 ÷ 1000 = 0.341

23 500 ÷ 10 000 = 2.35

You always move the DECIMAL POINT this much:
1 place for 10, 2 places for 100,
3 places for 1000, 4 for 10 000 etc.

4) To Divide by Numbers like 40, 300, 7000 etc.

DIVIDE by 4 or 3 or 7 etc. FIRST, then move the Decimal Point so many places SMALLER (i.e. to the left).

Example:

To find 960 ÷ 300, first divide by 3 — 960 ÷ 3 = 320,
then move the DP 2 places smaller — = 3.2

With love to guide us, nothing can divide us...

... a great man once said. In other news, knowing how to divide by multiples of 10 will be very handy.

1) Work out a) 2.45 ÷ 10 b) 654.2 ÷ 100 c) 3.08 ÷ 1000

2) Work out a) 32 ÷ 20 b) 360 ÷ 30 c) 4000 ÷ 800

Multiplying Without a Calculator

You need to be really happy doing multiplications without a calculator — you'll definitely need to do it in your exam. So make sure you learn the methods on this page...

Multiplying Whole Numbers

There are lots of methods you can use for this. Two of the popular ones are shown below. Just make sure you can do it using whichever method you prefer...

The Traditional Method:

Split it into separate multiplications, and then add up the results in columns (right to left).

```
     46
  ×  27
  -----
    322   ← This is 7 × 46
    920   ← This is 20 × 46
  -----
   1242
```

The Grid Method:

Split up the two numbers into their tens and units and arrange the calculation in a grid.
Then do 4 easy multiplications to fill up the grid...

6 × 20 = 120
40 × 20 = 800
6 × 7 = 42
40 × 7 = 280

	40	6	Totals:
20	800	120	→ 920
7	280	42	→ + 322
			1242

Finally, add up the 4 numbers in the grid to get the answer.

Answer in both cases: 46 × 27 = 1242

Multiplying Decimals

OK, this is a little more tricky — so you'll just have to make sure you learn it.

1) To start with, forget about the decimal points and do the multiplication using whole numbers.
(E.g. for 1.2 × 3.45 you'd do 12 × 345.)
2) Now count the total number of digits after the decimal points in the original numbers.
(E.g. 1.2 and 3.45 — so that's 3 digits after the decimal point.)
3) Make the answer have the same number of decimal places.

EXAMPLE: "Work out 4.6 × 2.7"

1) We know that 46 × 27 = 1242 ('cos we've just worked it out)
2) 4.6 × 2.7 has 2 digits after the decimal points
3) So the answer is 12.42

Dividing Without a Calculator

OK, time for dividing now. Just remember, if you don't learn these basic methods, you'll find yourself in real trouble in the exam...

Dividing Whole Numbers

EXAMPLE: "What is 896 ÷ 8?"

$$\begin{array}{r}1\\ 8\overline{)8\,9\,6}\end{array}$$

8 into 8 goes once.

$$\begin{array}{r}1\,1\\ 8\overline{)8\,9\,{}^{1}6}\end{array}$$

8 into 9 goes once, carry the remainder of 1.

$$\begin{array}{r}1\,1\,2\\ 8\overline{)8\,9\,{}^{1}6}\end{array}$$

8 into 16 goes twice, so 896 ÷ 8 = 112.

EXAMPLE: "What is 312 ÷ 24?"

$$\begin{array}{r}0\\ 24\overline{)3\,1\,2}\end{array}$$

24 won't go into 3 so put a 0.

$$\begin{array}{r}0\,1\\ 24\overline{)3\,1\,{}^{7}2}\end{array}$$

24 into 31 goes once, carry the remainder of 7.

$$\begin{array}{r}0\,1\,3\\ 24\overline{)3\,1\,{}^{7}2}\end{array}$$

24 into 72 goes three times, so 312 ÷ 24 = 13.

Dividing with Decimals

EXAMPLE: "What is 52.8 ÷ 3?"

Just set it out like the ones we've just done but put the decimal point in the answer right above the one in the question...

$$\begin{array}{r}1\\ 3\overline{)5\,{}^{2}2\,.\,8}\end{array}$$

3 into 5 goes once, carry the remainder of 2.

$$\begin{array}{r}1\,7\,.\\ 3\overline{)5\,{}^{2}2\,.\,{}^{1}8}\end{array}$$

3 into 22 goes 7 times, carry the remainder of 1.

$$\begin{array}{r}1\,7\,.\,6\\ 3\overline{)5\,{}^{2}2\,.\,{}^{1}8}\end{array}$$

3 into 18 goes 6 times exactly. So 52.8 ÷ 3 = 17.6.

EXAMPLE: "What is 83.6 ÷ 0.4?"

The trick with ones like this is to remember it's a fraction: $\frac{83.6}{0.4}$

Now you can get rid of the decimals by multiplying the top and bottom by 10 (turning it into an equivalent fraction): $\frac{83.6}{0.4} = \frac{836}{4}$

It's now a lovely decimal-free division that you know how to solve:

$$\begin{array}{r}2\\ 4\overline{)8\,3\,6}\end{array}$$

4 into 8 goes twice.

$$\begin{array}{r}2\,0\\ 4\overline{)8\,3\,{}^{3}6}\end{array}$$

4 into 3 won't go so carry 3.

$$\begin{array}{r}2\,0\,9\\ 4\overline{)8\,3\,{}^{3}6}\end{array}$$

4 into 36 goes 9 times so 83.6 ÷ 0.4 = 209.

Sums without calculators — less painful than standing on a plug barefoot...

Hmm, lots of info to take in there — it's really important that you learn all the methods on these two pages though. Have another read if you're still a bit unsure. Then try all of these without a calculator:

1) 28 × 12	2) 56 × 11	3) 104 × 8
4) 96 ÷ 8	5) 242 ÷ 2	6) 357 ÷ 17
7) 3.2 × 56	8) 0.6 × 10.2	9) 5.5 × 10.2
10) 33.6 ÷ 0.6	11) 69 ÷ 1.5	12) 43.2 ÷ 3.6

Powers

Right. LEARN THIS PAGE, then PEEL ME SOME GRAPES and IRON MY SHIRTS.
Sorry, I think the power's gone to my head...

Powers are a very Useful Shorthand

$2\times2\times2\times2\times2\times2\times2 = 2^7$ ('two to the power 7')

$7\times7 = 7^2$ ('7 squared')

$6\times6\times6\times6\times6 = 6^5$ ('Six to the power 5')

$4\times4\times4 = 4^3$ ('four cubed')

That bit is easy to remember. Unfortunately, there are SIX SPECIAL RULES for powers that are not quite so easy, but you do need to know them for the Exam:

The Six Rules

The first two only work for powers of the SAME NUMBER.

1) When MULTIPLYING, you ADD the powers.

e.g. $3^4 \times 3^6 = 3^{4+6} = 3^{10}$ $\quad 8^3 \times 8^5 = 8^{3+5} = 8^8$

2) When DIVIDING, you SUBTRACT the powers.

e.g. $5^4 \div 5^2 = 5^{4-2} = 5^2$ $\quad 12^8/12^3 = 12^{8-3} = 12^5$

3) When RAISING one power to another, you MULTIPLY the powers.

e.g. $(3^2)^4 = 3^{2\times4} = 3^8$, $(5^4)^6 = 5^{24}$

4) $X^1 = X$, ANYTHING TO THE POWER 1 is just ITSELF

e.g. $3^1 = 3$, $\quad 6 \times 6^3 = 6^4$, $\quad 4^3 \div 4^2 = 4^{3-2} = 4^1 = 4$

5) $X^0 = 1$, ANYTHING TO THE POWER 0 is just 1

e.g. $5^0 = 1$ $\quad 67^0 = 1$ $\quad 3^4/3^4 = 3^{4-4} = 3^0 = 1$

6) $1^x = 1$, 1 TO ANY POWER is still just 1

e.g. $1^{23} = 1$ $\quad 1^{89} = 1$ $\quad 1^2 = 1$ $\quad 1^{1012} = 1$

"I've got the power! oh, oh, oh, oh..." Oh no, I'm feeling kinda nineties...

Some say you should learn the Six Rules for Powers. Then turn over and write it all down. Well, I say. Keep trying until you can do it! Then cover the page and apply the rules to simplify these.

1) a) $3^2 \times 3^6$ b) $4^3 \div 4^2$ c) $(8^3)^4$ d) $(3^2 \times 3^3 \times 1^6) / 3^5$ e) $7^3 \times 7 \times 7^2$

2) a) $5^2 \times 5^7 \times 5^3$ b) $1^3 \times 5^0 \times 6^2$ c) $(2^5 \times 2 \times 2^6) \div (2^3 \times 2^4)$

Square Roots and Cube Roots

Take a deep breath, and get ready to tackle this page. Good luck with it, I'll be rootin' for ya...

Square Roots

'Squared' means 'times by itself' : $P^2 = P \times P$
— SQUARE ROOT is the reverse process.

The best way to think of it is this:

'Square Root' means 'What Number Times by Itself gives...'

Example: 'Find the square root of 49' (i.e. 'Find $\sqrt{49}$')
To do this you should say "what number times by itself gives... 49?"
And the answer of course is 7.

Square Roots can be Positive or Negative

When you take the square root of a number, the answer can actually be positive or negative... you always have a positive and negative version of the same number.

E.g. $x^2 = 4$ gives $x = \pm\sqrt{4} = +2$ or -2

To understand why, look at what happens when you work backwards by squaring the answers: $2^2 = 2 \times 2 = 4$ but also $(-2)^2 = (-2) \times (-2) = 4$

Cube Roots

'Cubed' means 'times by itself twice' : $T^3 = T \times T \times T$ (once, twice)
— CUBE ROOT is the reverse process.

'Cube Root' means 'What Number Times by Itself TWICE gives...'

Example: "Find the cube root of 64" (i.e 'Find $\sqrt[3]{64}$')

You should say "What number times by itself twice gives... 64?"
And after a few guesses, the answer is 4.
(Note — unlike square roots, there's only ever one answer.)

"Cue brute", that's what I call Charley when I play him at snooker...

LEARN the 2 statements in the blue boxes and the methods for finding the roots.
Then turn the page and write it all down.

1) Find: a) $\sqrt{81}$ b) $\sqrt[3]{27}$
If you got only one answer for a) what is the other answer?
2) a) If $g^2 = 36$, find g. b) If $b^3 = 64$, find b. c) If $4 \times r^2 = 36$, find r.

Fractions, Decimals and Percentages

The one word that could describe all these three is PROPORTION.

Fractions, decimals and percentages are simply three different ways of expressing a proportion of something — and it's pretty important you should see them as closely related and completely interchangeable with each other. This table shows the really common conversions which you should know straight off without having to work them out:

Fraction	Decimal	Percentage
1/2	0.5	50%
1/4	0.25	25%
3/4	0.75	75%
1/3	0.333333... or $0.\dot{3}$	$33\frac{1}{3}\%$
2/3	0.666666.... or $0.\dot{6}$	$66\frac{2}{3}\%$
1/10	0.1	10%
2/10	0.2	20%
X/10	0.X	X0%
1/5	0.2	20%
2/5	0.4	40%

⅓ and ⅔ have what're known as 'recurring' decimals — the same pattern of numbers carries on repeating itself forever. (Except here, the pattern's just a single 3 or a single 6. You could have, for instance: 0.143143143...)

The more of those conversions you learn, the better — but for those that you don't know, you must also learn how to convert between the three types. These are the methods:

Fraction —Divide→ **Decimal** —× by 100→ **Percentage**

e.g. $\frac{1}{2}$ is 1÷2 = 0.5 e.g. 0.5 × 100 = 50%

Fraction ←The tricky one— **Decimal** ←÷ by 100— **Percentage**

Converting decimals to fractions is only possible for exact decimals that haven't been rounded off. It's simple enough, but it's best illustrated by examples — see below. You should be able to work out what the simple rule is.

$0.6 = \frac{6}{10}$ $0.3 = \frac{3}{10}$ $0.7 = \frac{7}{10}$ $0.x = \frac{x}{10}$ etc. These can then be cancelled down.

$0.12 = \frac{12}{100}$ $0.78 = \frac{78}{100}$ $0.45 = \frac{45}{100}$ $0.05 = \frac{5}{100}$ etc.

$0.345 = \frac{345}{1000}$ $0.908 = \frac{908}{1000}$ $0.024 = \frac{24}{1000}$ $0.xyz = \frac{xyz}{1000}$ etc.

There's more on fractions, decimals and percentages on pages 8-11.

Make sure you learn the method for finding percentages without a calculator.

Recurring decimals like 0.3333... are all actually just exact fractions in disguise. But don't worry, you don't need to know how to convert them back into fractions.

Oh, what's recurrin'?...

Knowing all of the top table and the 4 conversion processes for FDP will speed you up nicely in exams.

1) Turn the following decimals into fractions and reduce them to their simplest form.

a) 0.6 b) 0.02 c) 0.77 d) 0.555 e) 5.6

Fractions

Here are some more ways to cope with fraction calculations without your beloved calculator. (For some tricks on doing fractions with your calculator, see p6.)

1) Multiplying — easy

Multiply top and bottom separately:

$$\frac{3}{5} \times \frac{4}{7} = \frac{3 \times 4}{5 \times 7} = \frac{12}{35}$$

2) Dividing — quite easy

Turn the 2nd fraction UPSIDE DOWN and then multiply:

$$\frac{3}{4} \div \frac{1}{3} = \frac{3}{4} \times \frac{3}{1} = \frac{3 \times 3}{4 \times 1} = \frac{9}{4}$$

3) Adding, Subtracting — fraught

Add or subtract TOP LINES ONLY but only if the denominators (bottom numbers) are the same.

(If they're not, you have to make them the same — see below).

$$\frac{2}{6} + \frac{1}{6} = \frac{3}{6}$$

$$\frac{5}{7} - \frac{3}{7} = \frac{2}{7}$$

Don't forget — you also need to be able to find 'fractions of' things. Have a look back at page 8 to remind yourself.

4) Ordering Fractions — Put them over the Same No.

E.g. Put these fractions in ascending order of size: $\frac{8}{3}, \frac{6}{4}, \frac{12}{5}$

① First, to find the new denominator, just find the LCM (see p39) of the denominators: LCM of 3, 4 and 5 is 60

② Then change each fraction so it's over the new number:

$$\frac{8}{3} \underset{\times 20}{\overset{\times 20}{=}} \frac{160}{60} \qquad \frac{6}{4} \underset{\times 15}{\overset{\times 15}{=}} \frac{90}{60} \qquad \frac{12}{5} \underset{\times 12}{\overset{\times 12}{=}} \frac{144}{60}$$

③ Now they're easy to write in order: $\frac{90}{60}, \frac{144}{60}, \frac{160}{60}$ or $\frac{6}{4}, \frac{12}{5}, \frac{8}{3}$.

5) Dealing with Mixed Numbers

There's only one way to deal with calculations involving mixed numbers (that's things like $3\frac{1}{3}$):

1) change them to 'normal' fractions
2) then you can do them as usual.

Don't try to do them any other way, or you'll probably get it wrong.

$$3\frac{2}{3} \times 7\frac{3}{4} = \frac{11}{3} \times \frac{31}{4} = \frac{341}{12}$$

($11 = 2 + (3 \times 3)$, $31 = 3 + (7 \times 4)$)

No fractions were harmed in the making of this page...

Now you've learnt those five useful tips, it's time to have a go at these below — without a calculator.

1) a) $\frac{3}{8} \times \frac{5}{12}$ b) $\frac{4}{5} \div \frac{7}{8}$ c) $\frac{3}{4} + \frac{2}{5}$ d) $\frac{2}{5} - \frac{3}{8}$ e) $4\frac{1}{9} + 2\frac{2}{27}$

2) Order the following fractions: $\frac{11}{15}, \frac{3}{5}, \frac{2}{3}$

Ratios

Ah ratios... You may remember these from Unit 1 (see p.12). In which case, you'll have no problem reciting the legendary GOLDEN RULE...

DIVIDE FOR ONE, THEN TIMES FOR ALL

Example 1: "Divide £400 in the ratio 5:3"

The GOLDEN RULE says: **DIVIDE FOR ONE, THEN TIMES FOR ALL**

The trick with this type of question is to add together the numbers in the RATIO to find how many PARTS there are: 5 + 3 = 8 parts.

Now use The Golden Rule:

Divide the £400 by 8 to find how much it is for ONE PART, then multiply by 5 and by 3 to find how much 5 PARTS ARE and how much 3 PARTS ARE.

So...

£400 ÷ 8 = £50 (for 1 part)
×5 = £250 (for 5 parts)
×3 = £150 (for 3 parts)

So £400 split in the ratio 5:3 is £250 : £150

Example 2: "The milkman delivers skimmed and full fat milk to a house in the ratio 3:2. If he delivers 6 bottles of full fat milk, how much skimmed milk does he deliver?"

The GOLDEN RULE says: **DIVIDE FOR ONE, THEN TIMES FOR ALL**

which means:

Divide the number of bottles of full fat by 2 to find how many bottles is ONE PART, then multiply by 3 to find how many bottles of skimmed are delivered.

So...

6 bottles ÷ 2 = 3 bottles (for one part)
× 3 = 9 bottles of skimmed (for 3 parts).

One rule to... er... rule them all...

You should know the rule off by heart now, but make sure you know when to use it too. Learn the examples above and put your skills to the test with these questions.

1) Chris and Steve have ages in the ratio 3:4. If Chris is 18, how old is Steve?
2) Divide £2400 in the ratio 5:7.

Revision Test for Unit 2 — 1

WHAT YOU'RE SUPPOSED TO DO HERE is put all the methods of this Unit so far into practice to answer these questions.

DON'T use your calculator for these questions.

1) Write this number out in words: 21 306 515

2) Put these numbers in order of size:
23 6534 123 2200 2 132 789 45

3) Jan needs to buy 100 large envelopes at £1.20 each. How much money does she need?

4) Chris bought 100 chocolate bars for £32 to sell in the school tuck shop.
How much should he sell each chocolate bar for to make his money back?

5) It cost Asik £150 to print 300 programmes for the School play.
How much should he sell each programme for to make his money back?

6) Find all the prime numbers between 40 and 60 (there are 5 of them).

7) What are Multiples? Find the first six multiples of 10 and also of 4.

8) What are Factors? Find all the factors of 30.

9) Express as a product of prime factors: a) 210 b) 1050

10) Find the highest common factor of 42 and 28.

11) Find the lowest common multiple of 8 and 10.

12) What are cube numbers? Write down the first ten of them.

13) Work out the value of: a) 6^1 b) 1^5 c) 4 cubed
d) 12^0 e) $2^3 \times 2^2$ f) 8 squared

14) Christine works in a biology lab. She has approximately 12^5 bacteria in some liquid in a test tube. She splits the liquid equally into 12 new test tubes. Approximately how many bacteria will there be in each test tube now?

15) Without a calculator, find all possible answers of a) $\sqrt{225}$ b) $\sqrt[3]{216}$.

16) Convert 0.645 into a fraction.

17) Work out without a calculator: a) $\frac{4}{6} \times 2\frac{12}{5}$ b) $\frac{25}{6} \div \frac{8}{3}$ c) $\frac{5}{8} + \frac{9}{4}$ d) $\frac{2}{3} - \frac{1}{7}$

18) What is The Golden Rule for Ratios?

19) Sarah is ordering new stock for her clothes shop. The shop usually sells red and blue scarves in the ratio 5:8. Sarah orders 150 red scarves. How many blue scarves should she order?

Negative Numbers and Letters

Algebra — the very word is enough to strike fear into the heart of even the strongest person. But fear ye not, for here I shall show you how to create a shield of light powerful enough to deflect even the strongest rays of algebra, at a distance of over 30 metres. I call it the mirror...

Rule 1

+	+	makes	+
+	–	makes	–
–	+	makes	–
–	–	makes	+

Only to be used when:

1) Multiplying or dividing

e.g. $-2 \times 3 = -6$, $-8 \div -2 = +4$ $-4p \times -2 = +8p$

2) Two signs appear next to each other

e.g. $5 - -4 = 5 + 4 = 9$ $4 + -6 - -7 = 4 - 6 + 7 = 5$

Rule 2

Use the NUMBER LINE for ADDING OR SUBTRACTING:

E.g. "Simplify $4X - 8X - 3X + 6X$"

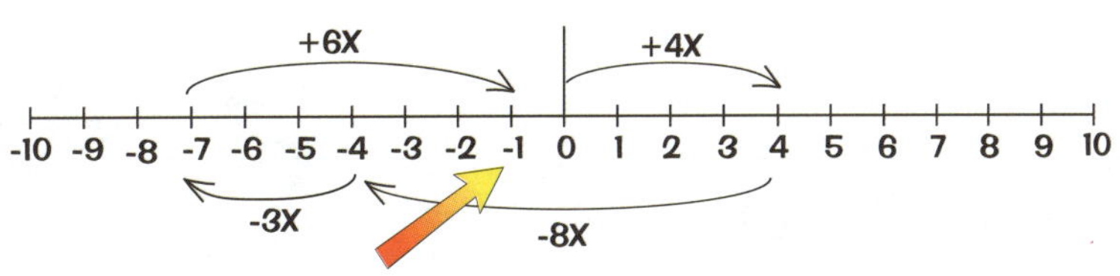

So $4X - 8X - 3X + 6X = -1X$

Letters Multiplied Together

This is the super-slick notation they like to use in algebra which just ends up making life difficult for folks like you. You've got to remember these five rules:

1) 'abc' means '$a \times b \times c$' — The ×'s are often left out to make it clearer.

2) 'gn^2' means '$g \times n \times n$' — Note that only the n is squared, not the g as well.

3) '$(gn)^2$' means '$g \times g \times n \times n$' — The brackets mean that BOTH letters are squared.

4) '$p(q - r)^3$' means '$p \times (q - r) \times (q - r) \times (q - r)$' — Only the brackets get cubed.

5) '-3^2' is too ambiguous. — It should either be written $(-3)^2 = 9$, or $-(3^2) = -9$.

Ahhh algebra, it's as easy as abc, or ab², or something like that...

Everyone knows Rule 1, but it can be easy to forget that sometimes you need Rule 2 instead. Make sure you know both rules and when to use them. You also need to learn the 5 special cases of Letters Multiplied Together.

1) For each of a) to d), decide where Rule 1 and Rule 2 apply, and then work them out.

a) -4×-3 b) $-4 + -5 + 3$ c) $(3X + -2X - 4X) \div (2 + -5)$ d) $120 \div -40$

2) If $m=2$ and $n=-3$, work out: a) mn^2 b) $(mn)^3$ c) $m(4+n)^2$ d) n^3 e) $3m^2n^3 + 2mn$

Algebra

Algebra really terrifies so many people. But honestly, it's not that bad. You've just got to make sure you understand and learn these basic rules for dealing with algebraic expressions. After that, all it needs is practice, practice, practice... and a little love.

1) Terms

Before you can do anything else, you MUST understand what a TERM is:

1) **A TERM IS A COLLECTION OF NUMBERS AND LETTERS, ALL MULTIPLIED/DIVIDED TOGETHER.**

2) Terms are SEPARATED BY + AND – SIGNS   e.g. $4x^2 - 3py - 5 + 3p$ (Terms)

3) Terms always have a + or – ATTACHED TO THE FRONT OF THEM

4) e.g. $(4xy)(+5x^2)(-2y)(+6y^2)(+4)$

Invisible + sign — 'xy' term — 'x^2' term — 'y' term — 'y^2' term — 'number' term

2) Simplifying

'Collecting Like Terms'

EXAMPLE: "Simplify $2x - 4 + 5x + 6$"

$(2x)(-4)(+5x)(+6) = (+2x)(+5x)(-4)(+6)$

x-terms   number terms $= 7x \quad +2 = \underline{7x + 2}$

1) Put bubbles round each term, — be sure you capture the +/– sign IN FRONT of each.
2) Then you can move the bubbles into the best order so that LIKE TERMS are together.
3) 'LIKE TERMS' have exactly the same combination of letters, e.g. 'x-terms' or 'xy-terms'.
4) Combine LIKE TERMS using the NUMBER LINE (not the other rule for negative numbers).

It could be worse. Probably. If you think about it. I'm sure it could...

Learn all the key facts on this page, then have a go at these questions to see how well you've got it.

1) Simplify: a) $5x + 3y - 4 - 2y - x$ b) $3x + 2 + 5xy + 6x - 7$
c) $2x + 3x^2 + 5y^2 + 3x$ d) $3y - 6xy + 3y + 2yx$

Algebra

3) Multiplying out Brackets

1) The thing OUTSIDE the brackets multiplies each separate term INSIDE the brackets.
2) When letters are multiplied together, they are just written next to each other, e.g. pq
3) Remember, $R \times R = R^2$, and TY^2 means $T \times Y \times Y$, whilst $(TY)^2$ means $T \times T \times Y \times Y$.
4) Remember a minus outside the bracket REVERSES ALL THE SIGNS when you multiply.

EXAMPLE: 1) $3(2x + 5) = 6x + 15$ 2) $4p(3r - 2t) = 12pr - 8pt$

3) $-4(3p^2 - 7q^3) = -12p^2 + 28q^3$ (note both signs have been reversed — Rule 4)

4) Factorising — putting brackets in

This is the exact reverse of multiplying-out brackets. Here's the method to follow:

1) Take out the biggest NUMBER that goes into all the terms.
2) For each letter in turn, take out the highest power (e.g. x, x^2 etc) that will go into EVERY term.
3) Open the brackets and fill in all the bits needed to reproduce each term.

EXAMPLE: Factorise $15x^4y + 20x^2y^3z - 35x^3yz^2$

ANSWER: $5x^2y(3x^2 + 4y^2z - 7xz^2)$

- Biggest number that'll divide into 15, 20 and 35
- Highest powers of x and y that will go into all three terms
- z wasn't in ALL terms so it can't come out as a common factor

REMEMBER: 1) The bits taken out and put at the front are the common factors.
2) The bits inside the brackets are what's needed to get back to the original terms if you were to multiply the brackets out again.

Go forth and multiply out brackets...

The techniques on these two pages really are the nuts n' bolts of algebra. There's no getting out of it, you'll have to knuckle down and learn everything on these pages inside out. Then have a crack at these:

1) Multiply out the brackets: a) $2(x - 2)$ b) $x(5 + x)$ c) $y(y + x)$ d) $3y(2x - 6)$

2) Factorise: a) $5xy + 15x$ b) $5a - 7ab$ c) $12xy + 6y - 36y^2$

Formulas

These can seem a bit confusing but they're not as bad as they look, once you know the "tricks of the trade", as it were. There are two main things you'll have to do with 'em...

Making Formulas from Words

There are two types of question they can ask. In the first type, they'll give you instructions about what to do with a number and you have to write it as a formula.
The only things they're likely to want you to do in the formula are:

1) Multiply x 2) Divide x 3) Add or subtract a number

Example 1: "To find y, multiply x by three and then subtract four"

Answer: Start with x → 3x → 3x – 4 so $y = 3x - 4$

Times it by 3, Subtract 4 (not too gruelling, is it?)

The second type are a bit harder. You have to make up a formula by putting in letters like 'C' for 'cost' or 'n' for 'number of something-or-others'.

Example 2: Froggatt's deep-fry 'CHOCCO-BURGERS' cost 58 pence each. Write a formula for the total cost, T, of buying n 'CHOCCO-BURGERS' at 58p each.

Answer: T stands for the total cost

n stands for the number of 'CHOCCO-BURGERS'

In words the formula is: Total Cost = Number of 'CHOCCO-BURGERS' × 58p

Putting the letters in: $T = n \times 58$ or to write it better: $T = 58n$

Substituting Values into Formulas

$$F = \frac{9}{5}C + 32$$

Example: Use the formula above to convert 15 °C from Celsius (C) into Fahrenheit (F).

Method:

1) Write out the Formula — e.g. $F = \frac{9}{5}C + 32$
2) Write it again, directly underneath, but substituting numbers for letters on the RHS (Right Hand Side). — $F = \frac{9}{5}15 + 32$
3) Work it out IN STAGES. Use BODMAS (see p.2) to work things out IN THE RIGHT ORDER. WRITE DOWN values for each bit as you go along. — $F = 27 + 32$, $= 59$, $F = 59°$
4) DO NOT attempt to do it all in one go on your calculator — it'll only end in tears.

I'm so impressed with this page that words fail me...

Nope. I got nothin'. Best just do these questions:

1) The value of y is found by taking x, multiplying it by five and then subtracting three. Write down a formula for y in terms of x.
2) If $C = \frac{5}{9}(F - 32)$, find the value of C when F = 77.

Solving Equations

The 'proper' way to solve equations is shown on the next page. In practice the 'proper' way can be pretty difficult so there's a lot to be said for the much easier methods shown below.

The drawback with these is that you can't always use them on very complicated equations. In most Exam questions though, they do just fine.

1) The 'Common Sense' Approach

The trick here is to realise that the unknown quantity 'x' is after all just a number and the 'equation' is just a cryptic clue to help you find it

Example: "Solve this equation: $3x + 4 = 46$"
(i.e. find what number x is)

Answer: This is what you should say to yourself:

> "Something + 4 = 46" hmm, so that 'something' must be 42.
>
> So that means $3x = 42$, which means '3 times something = 42'
>
> So it must be $42 \div 3$ which is 14 so $x = 14$

In other words don't think of it as algebra, but as 'Find the mystery number'.

2) The Trial and Error Method

This is a perfectly good method, and although it won't work every time, it usually does, especially if the answer is a whole number.

The big secret of trial and error methods is to find TWO OPPOSITE CASES and keep taking values IN BETWEEN them.

In other words, find a number that makes the Right Hand Side bigger, and then one that makes the Left Hand Side bigger, and then try values in between them.

Example: "Solve for x: $3x + 5 = 21 - 5x$"
(i.e. find the number x)

Answer:

> Try x=1: $3 + 5 = 21 - 5$, $8 = 16$ — no good, RHS too big
>
> Try x=3: $9 + 5 = 21 - 15$, $14 = 6$ — no good, now LHS too big

SO TRY IN BETWEEN: x = 2: $6 + 5 = 21 - 10$, $11 = 11$, YES, so $x = 2$.

But I'm innocent your honour — this trial is an error...

~~Loan those two methods~~. No. ~~Learn these to meatheads~~. No. Learn these two methods. Splendid.

1) Solve: $4x - 12 = 20$
2) Solve: $3x + 5 = 5x - 9$

Solving Equations

The 'proper' way of solving equations isn't hard, it just needs lots of practice.

3) The 'Proper' Way

Golden Rules

1) Always do the SAME thing to both sides of the equation.
2) To get rid of something, do the opposite.
   The opposite of + is – and the opposite of – is +.
   The opposite of × is ÷ and the opposite of ÷ is ×.
3) Keep going until you have a letter on its own.

Example 1: Solve $5x = 15$

$5x = 15$ ← 5x means $5 \times x$, so do the opposite — divide both sides by 5.

$\underline{x = 3}$

Example 2: Solve $\frac{p}{3} = 2$

$\frac{p}{3} = 2$ ← $\frac{p}{3}$ means $p \div 3$, so do the opposite — multiply both sides by 3.

$\underline{p = 6}$

Example 3: Solve $4y - 3 = 17$

$4y - 3 = 17$ ← The opposite of –3 is +3 so add 3 to both sides.

$4y = 20$ ← The opposite of ×4 is ÷4 so divide both sides by 4.

$\underline{y = 5}$

Example 4: Solve $2(x + 3) = 11$

$2(x + 3) = 11$ ← The opposite of ×2 is ÷2 so divide both sides by 2.

$x + 3 = 5.5$ ← The opposite of +3 is –3 so subtract 3 from both sides.

$\underline{x = 2.5}$

Example 5: Solve $3x + 5 = 5x + 1$

there are x's on both sides, so subtract 3x from both sides. → $3x + 5 = 5x + 1$

the opposite of +1 is –1, so subtract 1 from each side. → $5 = 2x + 1$

the opposite of ×2 is ÷2, so divide each side by 2. → $4 = 2x$

$\underline{2 = x}$

4) Rearranging Formulas

You do this in exactly the same way that you solve equations — watch...

Example 6: Rearrange the formula $q = 3p + 4$ to make p the subject:

$q = 3p + 4$

The opposite of +4 is –4 so take 4 from both sides. → $q - 4 = 3p$

The opposite of ×3 is ÷3 so divide both sides by 3. → $\frac{q - 4}{3} = p$

Dear Jim — could you fix it for me to make P the subject of this formula...

It's always good to know the proper way to solve equations, just in case you get thrown a curveball in the exam and they give you a real nightmare of an equation to solve.

1) Solve these equations: a) $3x + 1 = 13$ b) $q/_4 = 8$ c) $5y + 4 = 2y - 2$
2) Rearrange this formula to make b the subject: $2(b - 3) = a$

Inequalities

I'm not going to lie to you, some of this is a bit nasty. But I reckon you're up to it...

The 4 Inequality Symbols:

$>$ means '**Greater than**' $\geq$ means '**Greater than or equal to**'
$<$ means '**Less than**' $\leq$ means '**Less than or equal to**'

REMEMBER, the one at the **BIG** end is **BIGGEST**

so '$x > 4$' and '$4 < x$' **BOTH** say: '**x is greater than 4**'

Algebra With Inequalities — this is generally a bit tricky

The thing to remember here is that **inequalities are just like regular equations**:

$$5x < x + 2$$
$$5x = x + 2$$

in the sense that **all the normal rules for solving equations** (See P.51-52) **apply**...

...**BUT WITH ONE BIG EXCEPTION**:

Whenever you MULTIPLY OR DIVIDE BY A NEGATIVE NUMBER, you must FLIP THE INEQUALITY SIGN.

Example: "Solve $5x < 6x + 2$"

Answer: Subtract 6x from both sides: $5x - 6x < 2$

combining the x-terms gives: $-x < 2$

To get rid of the '–' in front of x you need to **divide both sides by -1** — but remember that means the '$<$' has to be flipped as well, which gives:

$x > -2$ i.e. '**x is greater than -2**' is the answer

(The $<$ has flipped around into a $>$, because we divided by a –ve number)

This answer, $x > -2$, can be displayed on a number line like this:

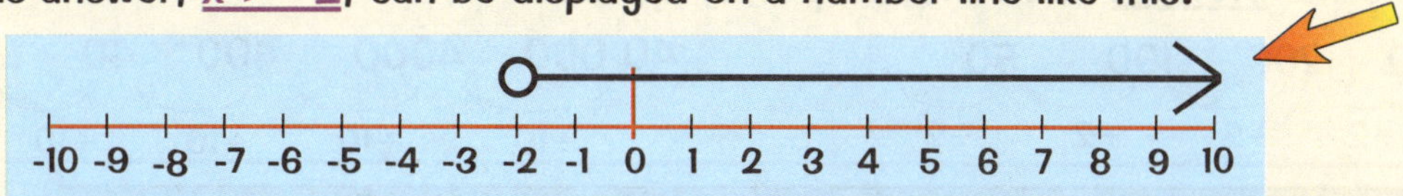

For $<$ and $>$, use an empty circle on the number line as shown — o.
For $\leq$ or $\geq$ use a solid circle — •.

The main thing you should realise, is that **MOST OF THE TIME** you just treat the '$<$' or '$>$' as though it was an '$=$' and **do all the usual algebra** that you would for a regular equation. The '**Big Exception**' doesn't actually come up very often at all.

Algebra with inequalities? Go and wash your mouth out young man...

Yikes, I'm glad that's over. Right, learn: The **4 Inequality Signs**, the **similarity** with **equations** and the **One Big Exception**. And then have a crack at these beautiful questions:

1) Solve this inequality: $4x + 3 \leq 6x + 7$.
2) Solve the inequalities and find the integer values of x which satisfy both: $2x + 9 \geq 1$ and $4x < 6 + x$

Number Patterns and Sequences

There are five different types of number sequences you could get in the exam, each as pretty as the last. They're not difficult — AS LONG AS YOU WRITE WHAT'S HAPPENING IN EACH GAP.

1) 'Add or Subtract the Same Number'

The SECRET is to write the differences in the gaps between each pair of numbers:

E.g. 2, 5, 8, 11, 14 ... (+3, +3, +3, +3, +3) 30, 24, 18, 12 ... (-6, -6, -6, -6)

The RULE: 'Add 3 to the previous term' 'Subtract 6 from the previous term'

2) 'Add or Subtract a Changing Number'

Again, WRITE THE CHANGE IN THE GAPS, as shown here:

E.g. 8, 11, 15, 20, 26 ... (+3, +4, +5, +6, +7) or 53, 43, 34, 26, 19 ... (-10, -9, -8, -7, -6)

The RULE: 'Add 1 extra each time to the previous term' 'Subtract 1 less from the previous term'

3) Multiply by the Same Number each Time

This type have a common MULTIPLIER linking each pair of numbers:

E.g. 5, 10, 20, 40 ... (×2, ×2, ×2, ×2) 2, 6, 18, 54 ... (×3, ×3, ×3, ×3)

The RULE: 'Multiply the previous term by 2' 'Multiply the previous term by 3'

4) Divide by the Same Number each Time

This type have the same DIVIDER between each pair of numbers:

E.g. 400, 200, 100, 50 ... (÷2, ÷2, ÷2, ÷2) 40 000, 4000, 400, 40 ... (÷10, ÷10, ÷10, ÷10)

The RULE: 'Divide the previous term by 2' 'Divide the previous term by 10'

5) Add the Previous Two Terms

This type of sequence works by adding the last two numbers to get the next one.

E.g. 1, 1, 2, 3, 5, 8, 13 ... (1+1, 1+2, 2+3, 3+5, 5+8, 8+13) 2, 4, 6, 10, 16 ... (2+4, 4+6, 6+10, 10+16)

The RULE: 'Add the previous two terms'

Number Patterns and Sequences

'State the Rule for Extending the Pattern'

This is what a lot of Exam questions end up asking for and it's easy enough so long as you remember this:

> **ALWAYS say what you do to the PREVIOUS TERM to get the next term.**

All the number sequences on the previous page have the rule for extending the pattern written in the box underneath them. Notice that they all refer to the previous term.

Finding the nth number:

You could be asked in the Exam to 'give an expression for the nth number in the sequence.' You'll only have to do this for a 'type 1' sequence (where the same number is added or subtracted). It's not that difficult because there's a simple formula:

'dn + (a – d)'

'd' is just THE DIFFERENCE between each pair of numbers.

'a' is just THE FIRST NUMBER in the sequence.

3, 7, 11, 15, 19, 23, 27 (differences: 4, 4, 4, 4, 4)

To get the nth term, you just find the values of 'a' and 'd' from the sequence and stick them in the formula. You don't replace n though — that wants to stay as n. Of course you have to learn the formula, but life is like that.

Example: "Find the nth number of this sequence: 5, 8, 11, 14 ....
and use it to find the 10th and 50th terms."

1) The formula is dn + (a – d)
2) The first number is 5, so a = 5. The differences are 3 so d = 3
3) Putting these in the formula gives: 3n + (5 – 3) = 3n + 2

So the nth number for this sequence is given by: '3n + 2'

4) Now you can use the formula to find any term in the sequence.
Just put the number of the term you want in place of n in the formula.
E.g.: 10th term: $3 \times 10 + 2 = 32$
50th term: $3 \times 50 + 2 = 152$

Look, if I've told you n times, I've told you n + 1 times — learn this page...

LEARN the 5 types of number patterns and the formula for finding the nth number and Bob's your uncle.

1) Find the next two numbers in each of these sequences, and say in words what the rule is for extending each one:
a) 2, 5, 9, 14 .... b) 2, 20, 200 .... c) 64, 32, 16, 8 ...

2) Find the expression for the nth number in this sequence: 7, 9, 11, 13

Coordinates

Graph questions can actually be quite fun. OK, maybe not fun exactly, but better than hopping blindfolded through a hail storm in just your pants. First, you need to get a grip of the basics...

The Four Quadrants

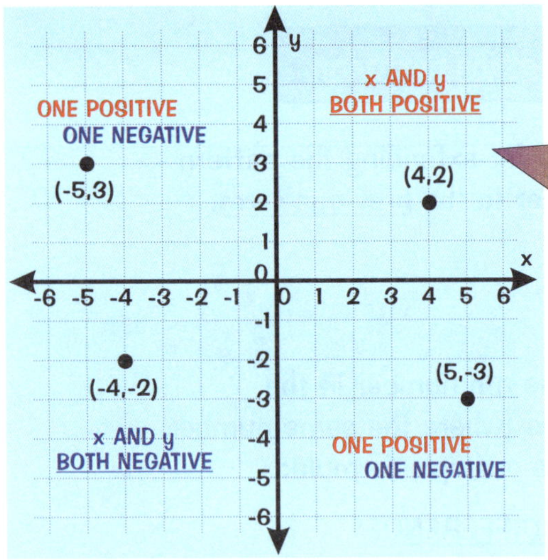

A graph has four different regions where the x- and y- coordinates are either positive or negative.

This is the easiest region by far because here ALL THE COORDINATES ARE POSITIVE.

You have to be dead careful in the OTHER REGIONS though, because the x- and y- coordinates could be negative, and that always makes life much more difficult.

X, Y Coordinates — Getting them in the Right Order

You must always give COORDINATES in brackets like this: (x, y)
And you always have to be real careful to get them the right way round, x first, then y.

Here are THREE POINTS to help you remember:

(x, y)

1) The two coordinates are always in ALPHABETICAL ORDER, x then y.
2) x is always the flat axis going ACROSS the page.
   In other words 'x is a..cross' Get it — x is a '×'. (Hilarious isn't it)
3) Remember it's always IN THE HOUSE (→) and then UP THE STAIRS (↑) so it's ALONG first and then UP, i.e. x-coordinate first, and then y-coordinate.

In the house and up the stairs? But I live in a bungalow...

Ah, a nice easy page this one. Make sure you learn the 3 Rules for getting x and y the right way round. Then turn over and write it all down.
Write down the coordinates of the letters A to H on this graph:

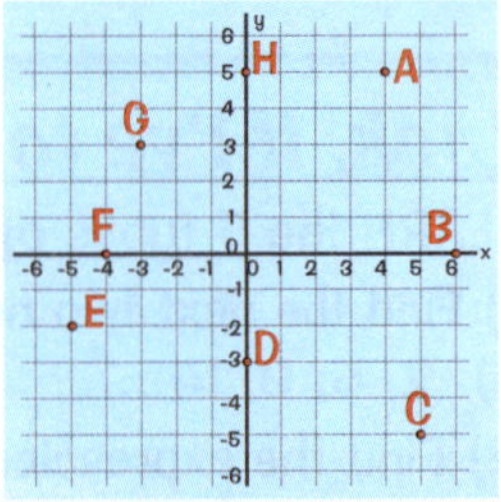

Straight Line Graphs — Gradients

Time to hit the slopes. Well, find them anyway...

Finding the Gradient

1) Find TWO ACCURATE POINTS

Both points should be in the upper right quadrant if possible (to keep all the numbers positive).

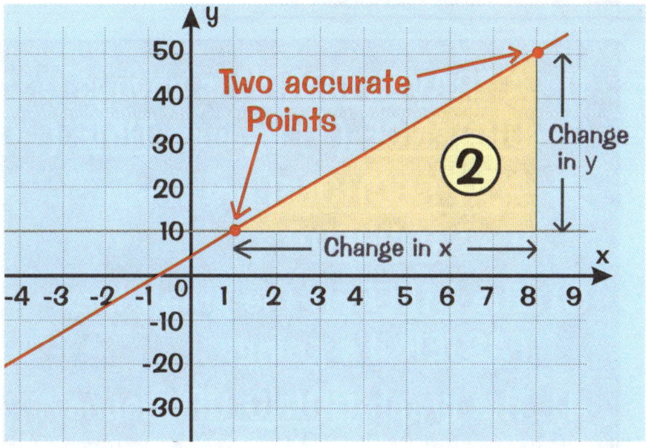

2) COMPLETE THE TRIANGLE as shown

3) Find the CHANGE IN Y and the CHANGE IN X

Make sure you subtract the x coords. the same way round as you do the y coords.
E.g. y coord. of pt A – y coord. of pt B
and x coord of pt A – x coord of pt B
Make sure you do this using the SCALES on the y- and x- axes, not by counting cm.
(So in the example shown, the change in y is 40 units from the y-axis.)

③ Change in y = 50 – 10 = 40
Change in x = 8 – 1 = 7

4) LEARN this formula, and use it:

$$\text{GRADIENT} = \frac{\text{CHANGE IN Y}}{\text{CHANGE IN X}} = \frac{\text{VERTICAL}}{\text{HORIZONTAL}}$$

④ Gradient $= \frac{40}{7} = 5.7$

Make sure you get it the right way up too! Remember it's VERy HOt — VERtical over HOrizontal

5) Finally, is the gradient POSITIVE or NEGATIVE?

If it slopes UPHILL left → right (↗) then it's positive
If it slopes DOWNHILL left → right (↘) then it's negative
(so put a minus(–) in front of it)

⑤ As the graph goes UPHILL, the gradient is positive. So 5.7 is correct, not -5.7

If you subtracted the coordinates the right way round, the sign should be correct. If it's not, go back and check what you've done.

Ch ch ch ch chaaaaaaaaaaaaaaaaaaaaaaaanges...

LEARN the FIVE STEPS for finding a gradient then turn over and WRITE THEM DOWN from memory.
Plot these 3 points on a graph: (0,3) (2,0) (5,-4.5) and then join them up with a straight line. Now carefully apply the FIVE STEPS to find the gradient of the line.

Travel and Conversion Graphs

Yay, just what you wanted to see — some more graphs...

Travel Graphs

1) A TRAVEL GRAPH is always DISTANCE (↑) against TIME (→)
2) FLAT SECTIONS are where it's STOPPED.
3) The STEEPER the graph the FASTER it's going.
4) The graph GOING UP means it's travelling AWAY. The graph COMING DOWN means it's COMING BACK AGAIN.

Travel Graph for a Very Tired cyclist

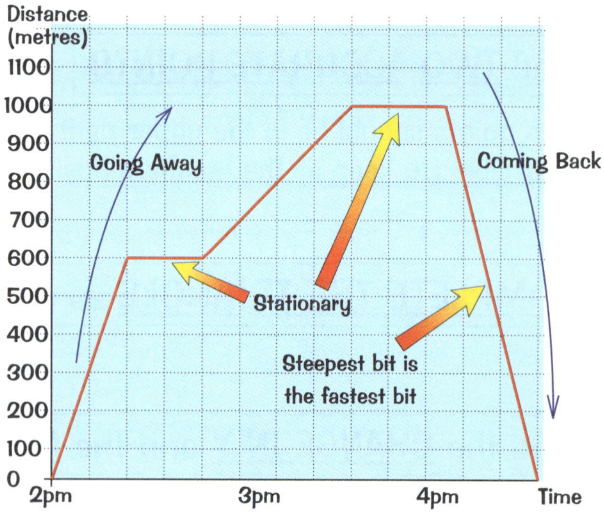

Conversion Graphs

These are really easy. In the Exam you're likely to get a Conversion Graph question which converts between things like £ → Dollars or mph → km/h, etc.

This graph converts between miles and kilometres

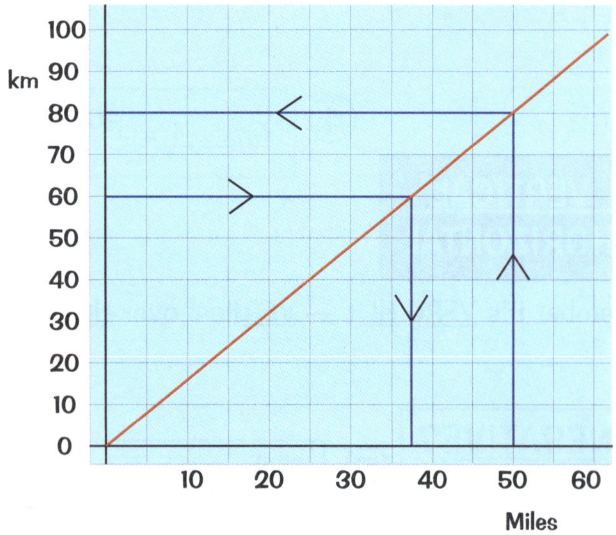

2 very typical questions:

1) How many miles is 60 km?

ANS: Draw a line straight across from '60' on the 'km' axis 'til it hits the line, then go straight down to the 'miles' axis and read off the answer: 37.5 miles

2) How many km is 50 miles?

ANS: Draw a line straight up from '50' on the 'miles' axis 'til it hits the line, then go straight across to the 'km' axis and read off the answer: 80 km

METHOD:

1) Draw a line from the value on one axis.
2) Keep going 'til you hit the LINE.
3) Then change direction and go straight to the other axis.
4) Read off the new value from the axis. That's the answer.

If you remember those 4 simple steps you really can't go wrong — let's face it, Conversion Graphs are a doddle.

Typical Graph Questions

It doesn't matter what type of graph you get asked questions about, distance vs time, chicken vs aliens — to read values from a graph you just need to follow the same easy method every time.

Getting Answers from your Graph

1) FOR A SINGLE CURVE OR LINE, you ALWAYS get the answer by drawing a straight line to the graph from one axis, and then down or across to the other axis, as shown here:

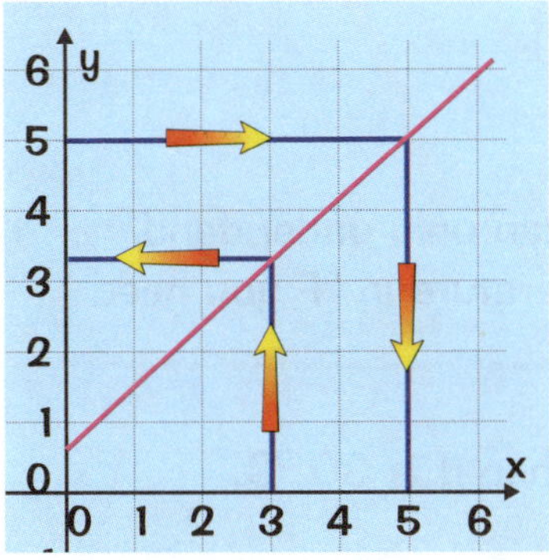

You should be fully expecting this to happen so that even if you don't understand the question, you can still have a pretty good stab at it:

If the question said "Find the value of y when x is equal to 3", ALL YOU DO IS THIS: start at 3 on the x-axis, go straight up to the graph, then straight over to the y-axis and read off the value, which in this case is y = 3.3 (as shown opposite).

2) IF TWO LINES CROSS.....
you can bet your very last fruitcake the answer to one of the questions will simply be:

THE VALUES OF X AND Y WHERE THEY CROSS
and you should be expecting that before they even ask it!

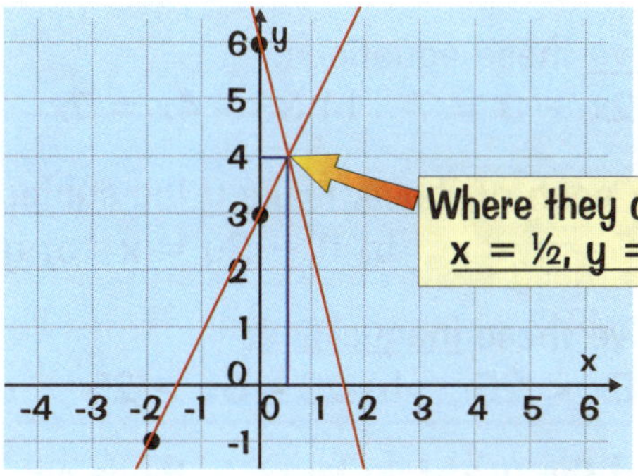

When the Graph is a Curve

The method even works when you don't have a straight line graph, as this example shows:

Example: "The graph shows the height of a ball as it is thrown through the air. Use the graph to find the approximate times when the ball was 1 m above the ground."

Method:

Draw a straight line across the graph at 1.0 on the y axis — this is where the height of the ball is 1 m.

The curve touches this line twice. Draw lines down to the x axis at these points and read off the values:

Time = 0.2 s and 0.6 s.

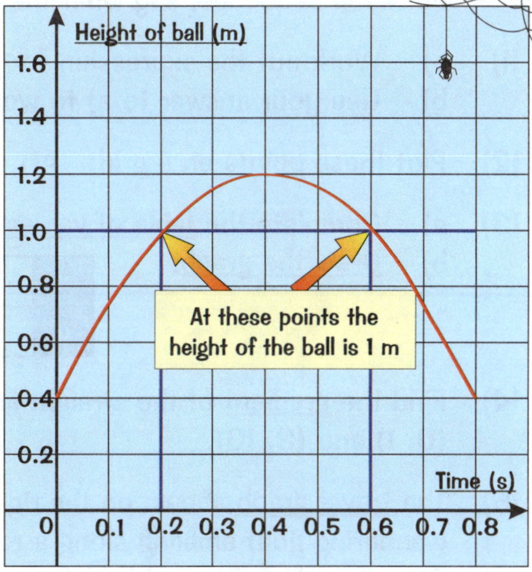

Right graph — we're not leaving here 'til I get some answers...

Learn the 2 simple Rules for getting answers.
Then turn over... bla bla bla ... you know what to do...

Revision Test for Unit 2 — 2

WHAT YOU'RE SUPPOSED TO DO HERE is put all the methods of the second half of this Unit into practice to answer these questions.

1) Work out: a) -3×-2 b) -4×8 c) $12 \div -4$ d) $-20 \div -4$

2) Simplify the expression: $3x + 4y + 2x - 4y$

3) Multiply out the brackets in the expression: $4(3g + 5h - 1)$

4) Factorise (take out common factors) this expression: $2x + 6xy$.

5) 'To find y you double x and add 4.' Write this as a formula.

6) Ellen's thermometer only measures in Celsius (°C) but her dad can only understand temperatures in Fahrenheit (°F). She knows that to get a temperature in °F you need to 'times the temperature in °C by 9/5 and then add 32'.
   a) Write out a formula for changing °C into °F.
   b) Use your formula to find the temperature for Ellen's dad when it is 30 °C.

7) Solve these equations:
   a) $2x + 3 = 7$ b) $33 - 4x = 7x$ c) $5(x + 5) = -10$

8) For each of these, make y the subject:
   a) $6 - y = x$ b) $11 + 2y = x$ c) $y/3 = 7x + 3$

9) Solve these inequalities:
   a) $5x < 25$ b) $20 - 5x > 25$ c) $-6x < 30$ d) $10x > 170 - 7x$

10) Find the next two terms in these sequences of numbers:
   a) 3, 7, 11, 15, ... b) 36, 28, 21, 15, 10, ...
   c) 3, 6, 12, 24, ... d) 1200, 600, 300, ...
   e) 2, 5, 7, 12, 19, ...
   For each sequence, say what the rule is for extending the pattern.

11) a) Work out the expression for the nth term in this sequence: –1, 5, 11, 17, 23, ...
   b) Use your answer to a) to work out the 10th term in the sequence.

12) Plot these points on a grid: A(1,4), B(-4,2), C(-3,-5), D(5,-6), E(5,0)

13) a) Complete the table of values below for the equation: $y = x + 3$
   b) Draw the graph.

x	-5	-3	-1	0	1	2	4	6
y	-2				4			

14) Find the gradient of the straight line that passes through the points (0, 1) and (3, 10).

15) The travel graph shown on the right relates to a wandering goat ambling along a road.
   Describe what the goat is doing
   a) Between 4 pm and 6 pm
   b) Between 6 pm and 7 pm
   c) How far from home does the goat stray?

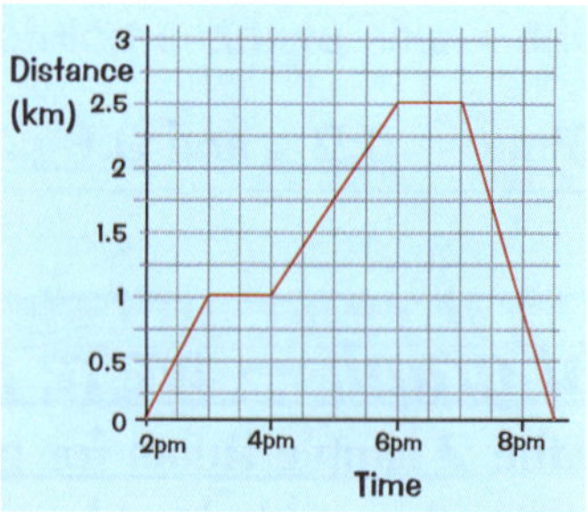

Trial and Improvement

This is a good method for finding approximate answers to equations that don't have simple whole number answers. Although it basically boils down to trial and error, there is a clear method which you must learn if you want to get it right...

You'll need to know about rounding to decimal places and significant figures for this — covered on p.4-5.

Method

1) SUBSTITUTE TWO INITIAL VALUES into the equation that give OPPOSITE CASES.
2) Choose your next value IN BETWEEN the previous two, and PUT IT into the equation.
3) After only 3 or 4 steps you should have 2 NUMBERS which are to the right degree of accuracy but DIFFER BY 1 IN THE LAST DIGIT.
4) Now take the EXACT MIDDLE VALUE to decide which is the answer you want.

Opposite cases means one answer too big, one too small. If they don't give opposite cases try again.

Continue the process, choosing new values between the two closest opposite cases, (and preferably nearer to the one which is closer to the answer you want).

E.g. if you had to get your answer to 2 DP, you'd eventually end up with say 5.43 and 5.44, with these giving OPPOSITE cases.

E.g. for 5.43 and 5.44, you'd try 5.435 to see if the real answer was between 5.43 and 5.435 or between 5.435 and 5.44.

Example

"The equation $x^3 + x = 40$ has a solution between 3 and 3.5. Find this solution to 1 DP"

Try $x = 3$ $\quad 3^3 + 3 = 30$ (Too small)
Try $x = 3.5$ $\quad 3.5^3 + 3.5 = 46.375$ (Too big)

← (2 opposite cases)

40 is what we want and it's closer to 46.375 than it is to 30 so we'll choose our next value for x closer to 3.5 than 3.

Try $x = 3.3$ $\quad 3.3^3 + 3.3 = 39.237$ (Too small)

Good, this is very close, but we need to see if 3.4 is still too big or too small:

Try $x = 3.4$ $\quad 3.4^3 + 3.4 = 42.704$ (Too big)

Good, now we know that the answer must be between 3.3 and 3.4.
To find out which one it's nearest to, we have to try the EXACT MIDDLE VALUE: 3.35

Try $x = 3.35$ $\quad 3.35^3 + 3.35 = 40.945$ (Too big)

This tells us with certainty that the solution must be between 3.3 (too small) and 3.35 (too big), and so to 1 DP it must round down to 3.3. ANSWER = 3.3

If at first you don't succeed — improve and trial again...

To get this method nailed, you must learn the 4 steps above. Do it now, and practise until you can write them down without having to look back at them. I said now. I can still see you — do it now!
The equation $x^3 - 2x = 1$ has a solution between 1 and 2. Find it to 1 DP.

Quadratic Graphs

Equations with an x^2 term in them are called quadratic equations. The graphs of these equations always have the same SYMMETRICAL bucket shape.

If the x^2 bit is positive (i.e. $+x^2$) the bucket is the normal way up, but if the x^2 bit has a "minus" in front of it (i.e. $-x^2$) then the bucket is upside down. The graphs get steeper and steeper but never vertical — remember this when you're drawing them.

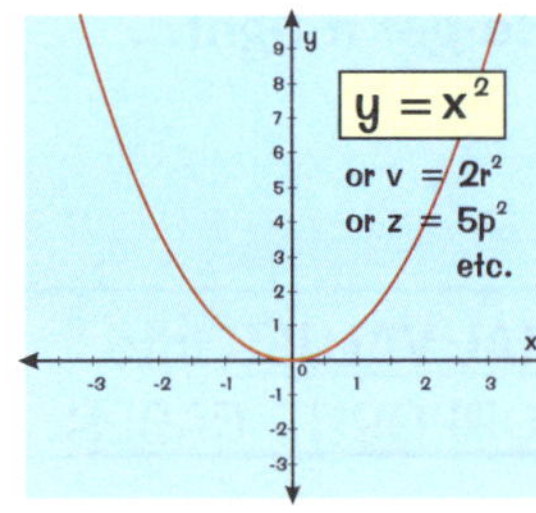

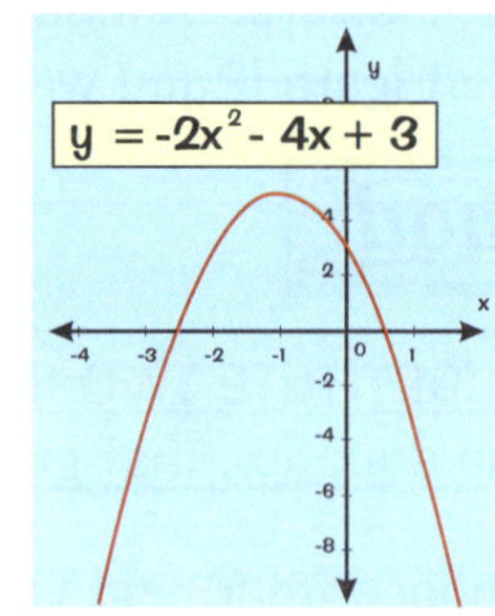

Most questions follow a set pattern...

Fill in the Table of Values

Example: "Fill in the table of values for the equation $y = x^2 + 2x - 3$ and draw the graph."

x	-5	-4	-3	-2	-1	0	1	2	3
y		5		-3	-4	-3	0		

Work out each point very carefully, writing down all your working. Don't just plug it all straight in your calculator — you'll make mistakes. To check you're doing it right, make sure you can reproduce the y-values they've already given you.

Draw the Curve

1) PLOT THE POINTS CAREFULLY, and don't mix up the x and y values.
2) The points should form a COMPLETELY SMOOTH CURVE. If they don't, they're wrong.

NEVER EVER let one point drag your line off in some ridiculous direction. When a graph is generated from an equation, you never get spikes or lumps — only MISTAKES.

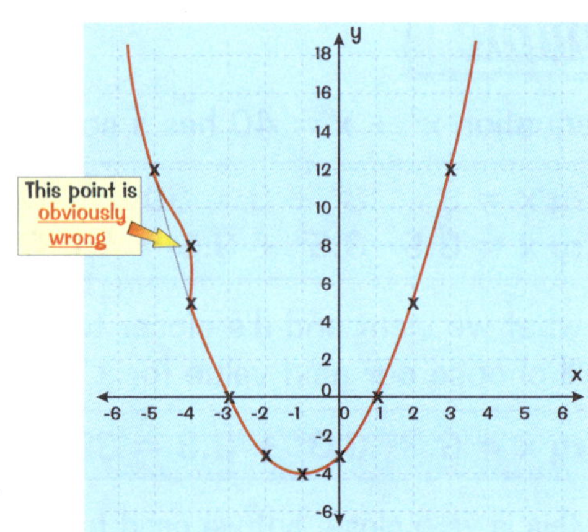

Use the Graph to Answer a Question

Example: "Use your graph to find the approximate values of x when $y = 0$ for the equation $y = x^2 + 2x - 3$."

1) To answer this question, all you do is read the x-values where $y = 0$, i.e. where it crosses the x-axis.
2) So the solutions are $x = -3$ and $x = +1$. (Quadratic equations usually have 2 solutions.)

Make sure your graphs are smoother than a polished kipper...

Learn the details of the method above for drawing quadratic graphs and finding approximate values of x and y by reading off the graph.

Plot the graph of $y = x^2 - x - 6$ (use x-values from -4 to 5).

Use your graph to find the approximate values of x when $y = 0$ for the equation $y = x^2 - x - 6$.

Symmetry

Symmetry is where a shape or picture can be put in different positions that look exactly the same. There are TWO types of symmetry you need to know about...

1) Line Symmetry

This is where you can draw a MIRROR LINE (or more than one) across a picture and both sides will fold exactly together.

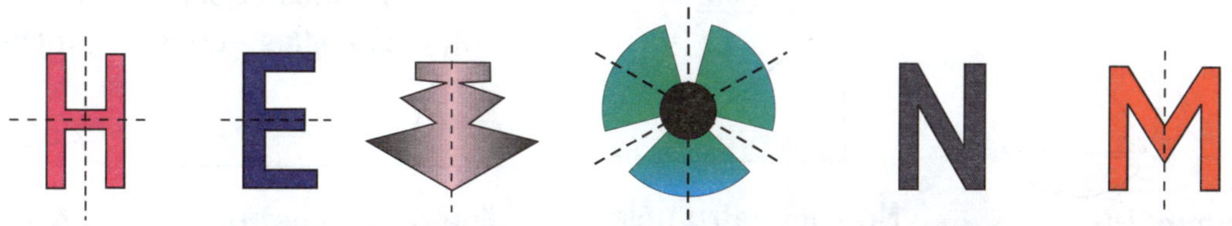

2 LINES OF SYMMETRY | 1 LINE OF SYMMETRY | 1 LINE OF SYMMETRY | 3 LINES OF SYMMETRY | NO LINES OF SYMMETRY | 1 LINE OF SYMMETRY

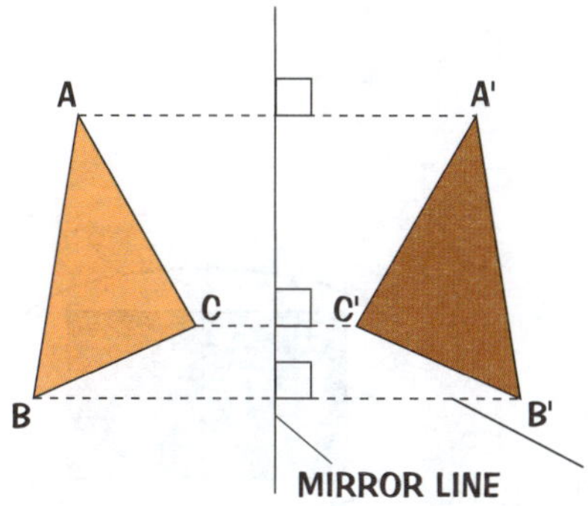

How to Draw a Reflection:

1) Reflect each point one by one.
2) Use a line which crosses the mirror line at 90° and goes EXACTLY the same distance on the other side of the mirror line, as shown.

A line which crosses at 90° is called a perpendicular

2) Rotational Symmetry

This is where you can ROTATE the shape or drawing into different positions that all look exactly the same.

Order 1 | Order 2 | Order 2 | Order 3 | Order 4

The ORDER OF ROTATIONAL SYMMETRY is the posh way of saying: 'HOW MANY DIFFERENT POSITIONS LOOK THE SAME'.
E.g. You should say the Z shape above has 'Rotational Symmetry order 2'

BUT... when a shape has ONLY 1 POSITION you can either say that it has 'Rotational Symmetry order 1' or that it has 'NO Rotational Symmetry'

I'd like to say I'm a fan of this topic, but I'm not very good at line...

Make sure you know the two types of symmetry, then copy the letters below and mark in all the lines of symmetry. Also say what the rotational symmetry is for each one.

H Z T N E X S

The Shapes You Need to Know

These are easy marks in the Exam — make sure you know them all.

Three-sided Shapes — Triangles

(just in case you didn't know...)

1) EQUILATERAL Triangle

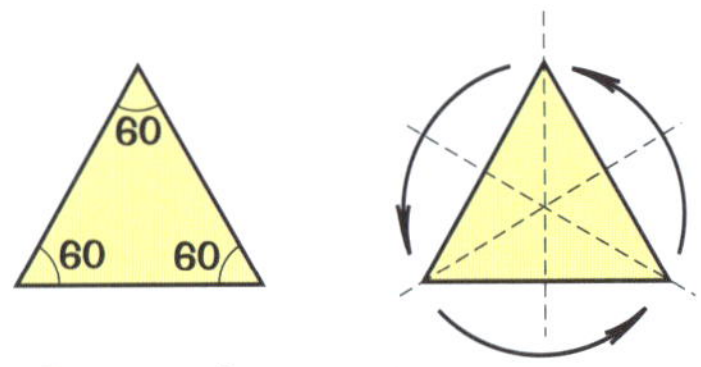

3 lines of symmetry.
Rotational symmetry order 3

2) RIGHT-ANGLED Triangle

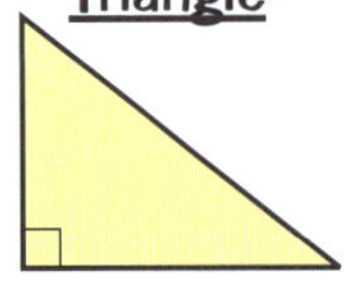

No symmetry unless the angles are 45°

3) ISOSCELES Triangle

2 sides equal
2 angles equal

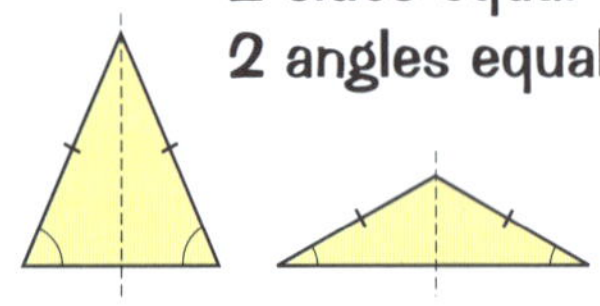

1 line of symmetry.
No rotational symmetry

4) SCALENE Triangle

No sides or angles equal

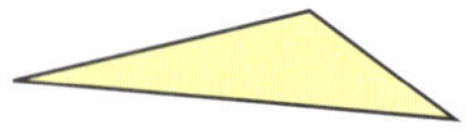

No symmetry.

Four-sided Shapes — Quadrilaterals

1) SQUARE

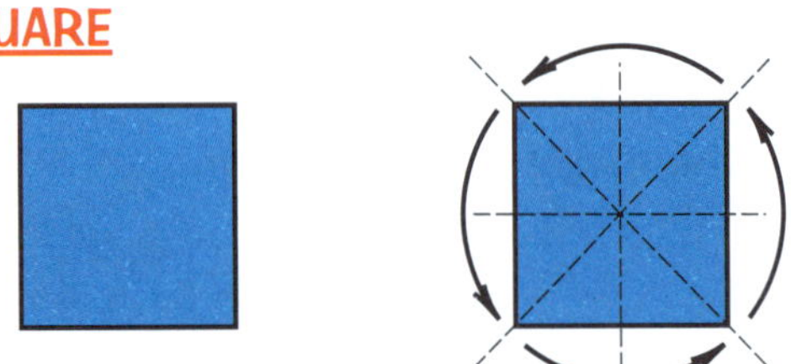

4 lines of symmetry.
Rotational symmetry order 4.

2) RECTANGLE

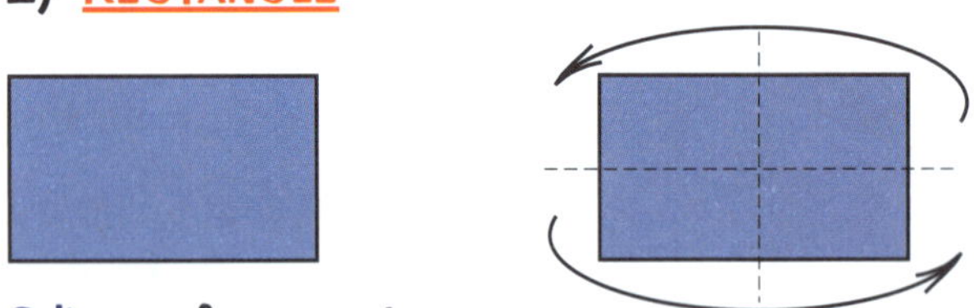

2 lines of symmetry.
Rotational symmetry order 2.

3) RHOMBUS (A square pushed over)
(It's also a diamond)

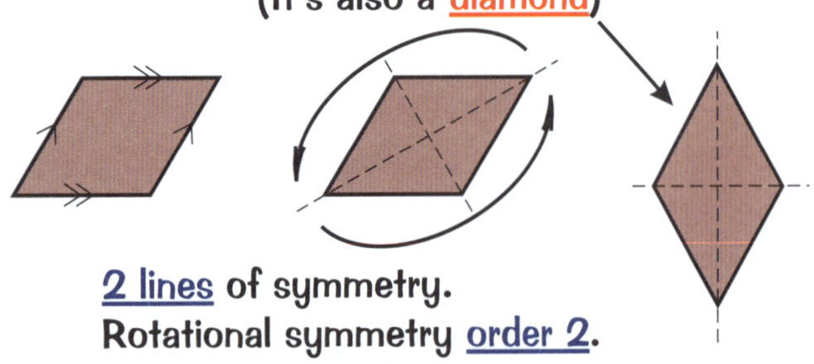

2 lines of symmetry.
Rotational symmetry order 2.

4) PARALLELOGRAM

(A rectangle pushed over — two pairs of parallel sides)

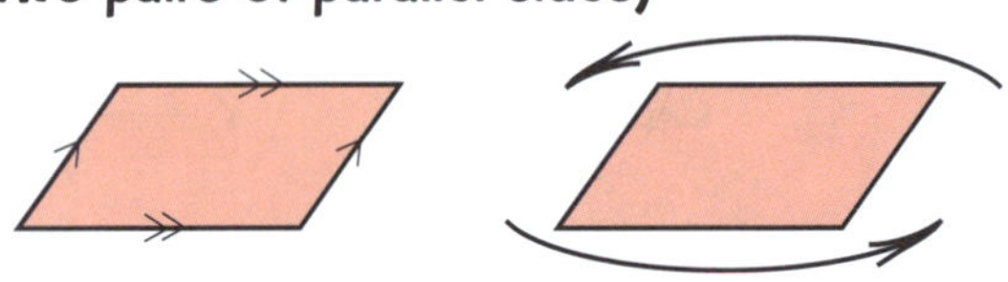

NO lines of symmetry.
Rotational symmetry order 2.

5) TRAPEZIUM (One pair of parallel sides)

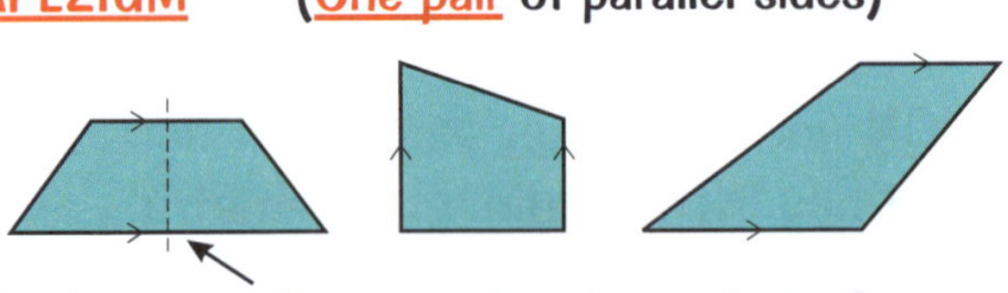

Only the isosceles trapezium has a line of symmetry.
None have rotational symmetry.

6) KITE

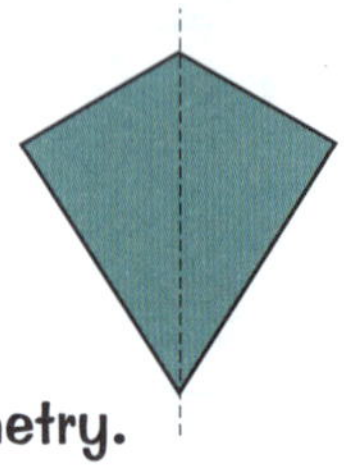

1 line of symmetry.
No rotational symmetry.

Rhombus facts: 4 sides, 2 lines of symmetry, Gemini, peanut allergy...

Learn everything on this page. Then turn over and write down all the details that you can remember. Then try again. It's as simple as that. Then you can play with the kite. Indoors though — it's new.

3D Shapes and Projections

I was going to make some pop-out 3D shapes to put on this page, but I couldn't find the scissors and sticky tape. Sorry. Still, you need to learn it all though — so chin up and learn the page.

Eight Solids To Learn

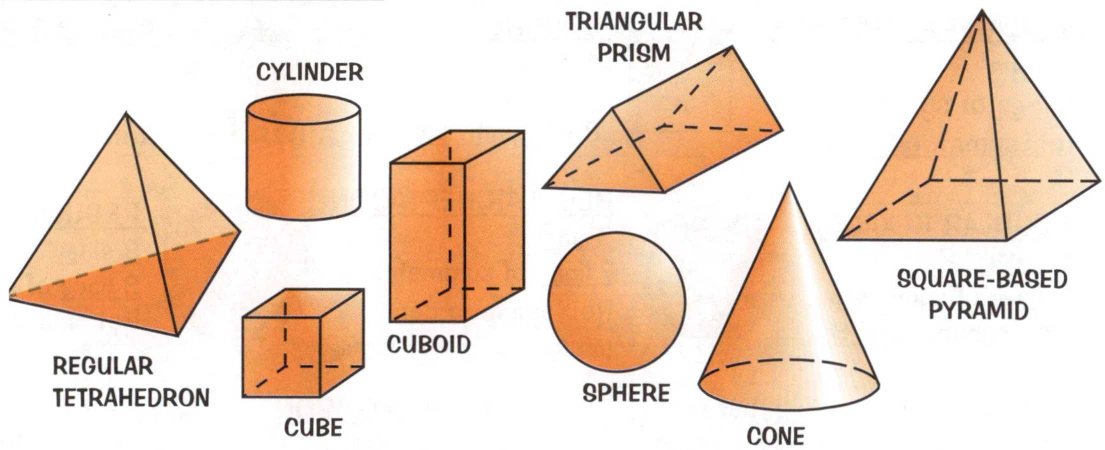

Projections show Different Views of a Shape

A 'projection' shows the relative size and shape of an object from either the front, side or back — they're usually known as 'elevations'. A 'plan' shows the view from above. They're always drawn to scale. Take this church (naff picture, I know) — you can represent it like this:

FRONT Elevation — the view you'd see if you looked from directly in front:

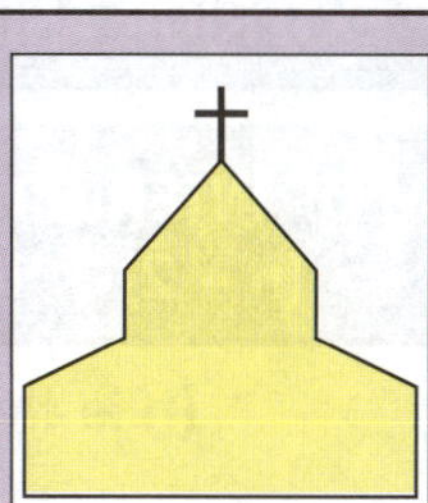

SIDE Elevation — the view you'd see if you looked from directly to one side:

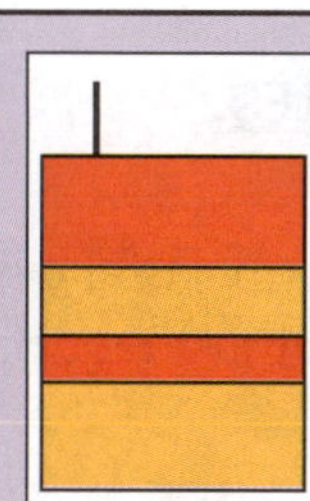

PLAN — the view you'd see if you looked from directly above:

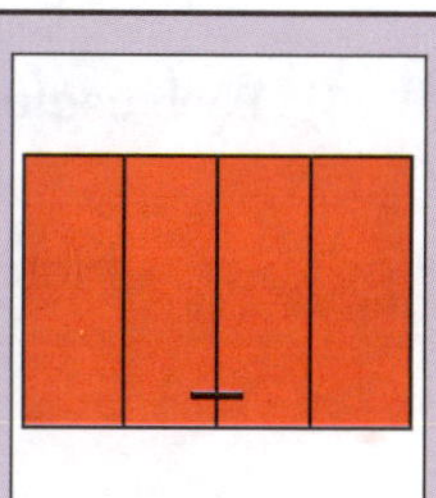

If they're feeling really mean (and they often are), you might get a question on:

This one's a bit trickier, so you might want to spend a little longer practising it — just to get your head round it.

ISOMETRIC Projection — this is where the shape is drawn (again, to scale) from a view at equal angles to all three axes (x, y and z). Or more simply, it's a drawing like this:

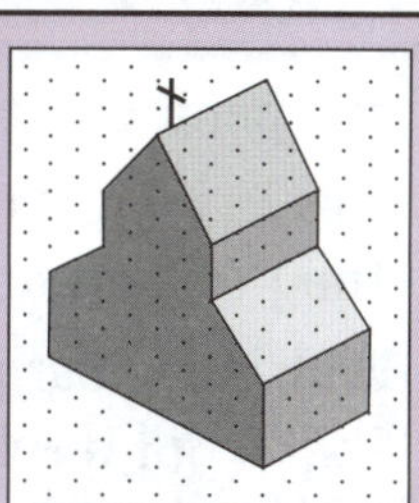

Well done — you've moved on to solids...

Deep breath in... and out... then learn the eight 3D shapes and the four types of projection. Cover up the page, then name 8 types of solid object.

1) Draw a plan, front and side elevations and an isometric projection of your own house.

Regular Polygons

A polygon is a many-sided shape. A regular polygon is one where all the sides and angles are the same. The regular polygons are a never-ending series of shapes with some fancy features.

You Need to Know These Polygons

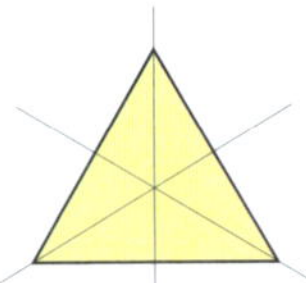

EQUILATERAL TRIANGLE
3 sides
3 lines of symmetry
Rot^nl symm. order 3

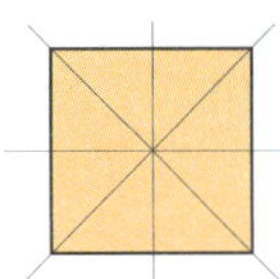

SQUARE
4 sides
4 lines of symmetry
Rot^nl symm. order 4

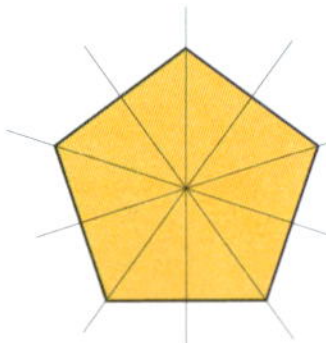

REGULAR PENTAGON
5 sides
5 lines of symmetry
Rot^nl symm. order 5

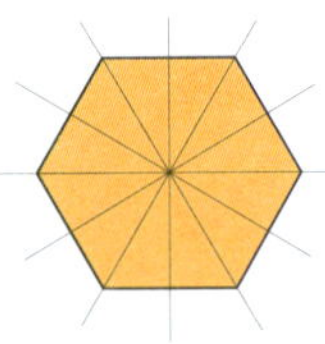

REGULAR HEXAGON
6 sides
6 lines of symmetry
Rot^nl symm. order 6

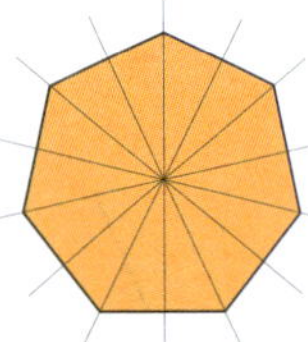

REGULAR HEPTAGON
7 sides
7 lines of symmetry
Rot^nl symm. order 7
(A 50p piece is like a heptagon)

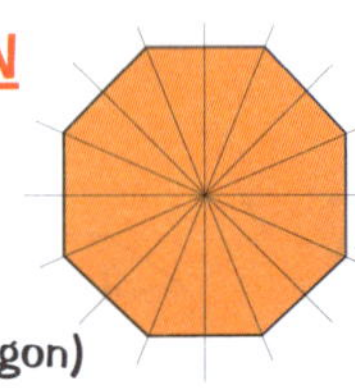

REGULAR OCTAGON
8 sides
8 lines of symmetry
Rot^nl symm. order 8

You also need to know the next two, but I'm not drawing them for you. Learn their names:

REGULAR NONAGON
9 sides, etc. etc.

REGULAR DECAGON
10 sides, etc. etc.

Polygons Have Interior and Exterior Angles

1) Exterior Angles

2) Interior Angles

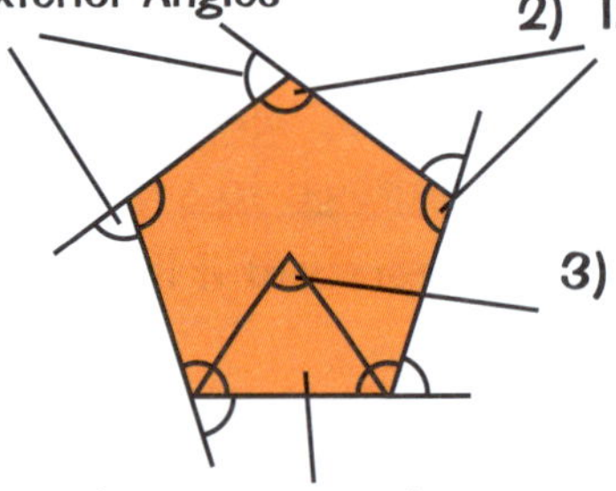

3) This angle is always the same as the Exterior Angles.

4) Each sector triangle is ISOSCELES.

There are 4 formulas to learn:

$$\text{EXTERIOR ANGLE} = \frac{360^\circ}{n}$$

$$\text{INTERIOR ANGLE} = 180^\circ - \text{EXTERIOR ANGLE}$$

$$\text{SUM OF EXTERIOR ANGLES} = 360^\circ$$

$$\text{SUM OF INTERIOR ANGLES} = (n - 2) \times 180^\circ$$

Note — the two SUM formulas above work for *any* polygons, not just regular ones.

(n is the number of sides).

Regular Polygons have Loads of Symmetry

1) The pentagon shown here has only 3 different angles in the whole diagram.
2) This is typical of regular polygons.
   They display an amazing amount of symmetry.
3) With a regular polygon, if two angles look the same, they will be.
   That's not a rule you should normally apply in geometry, and anyway you'll need to prove they're equal.

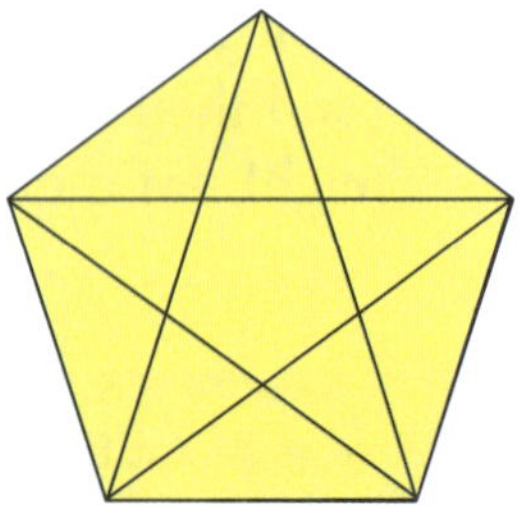

EXCLUSIVE: Heptagon lottery winner says "I'm still just a regular guy"...

Lots to learn on this page. But as long as you're familiar with the word regular, you're half way there.

1) What is a regular polygon?
2) Name the first six of them.
3) Work out the two key angles for a regular pentagon
4) And for a 12-sided regular polygon.

Perimeters

Perimeter is the distance all the way around the outside of a 2D shape. It's pretty straightforward to work out if you use the big blob method. So pay attention — this could be easy marks.

The Perimeter is the Distance Around the Edge of a Shape

To find a PERIMETER, you ADD UP THE LENGTHS OF ALL THE SIDES, but...
THE ONLY RELIABLE WAY to make sure you get all the sides is this:

1) PUT A BIG BLOB AT ONE CORNER and then go around the shape.
2) WRITE DOWN THE LENGTH OF EVERY SIDE as you go along.
3) EVEN SIDES THAT SEEM TO HAVE NO LENGTH GIVEN — you must work them out.
4) Keep going until you get back to the BIG BLOB.

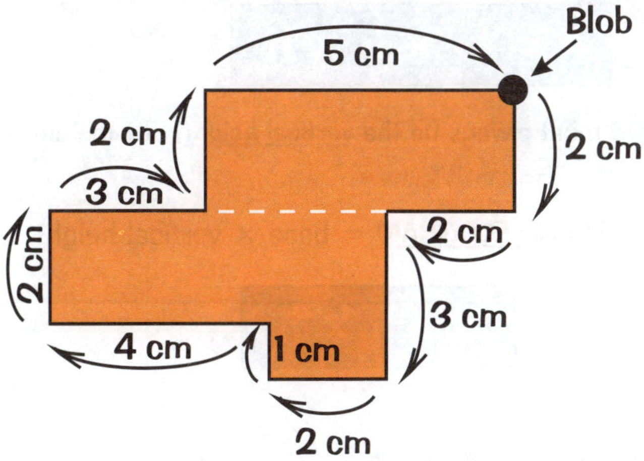

You must choose yourself a blob and it must also choose you. It will then be yours for life.

E.g. 2+2+3+2+1+4+2+3+2+5 = 26 cm

Yes, I know you think it's yet another fussy method, but believe me, it's so easy to miss a side. You must use good reliable methods for everything — or you'll lose marks willy nilly.

Finding the perimeter of a regular shape is dead easy — you just find the length of one side and multiply it by the number of sides.

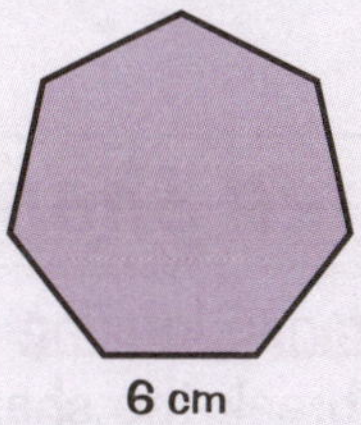

E.g. for this lovely heptagon, each side is 6 cm long.
There are 7 sides, so
Perimeter = 6 × 7 = 42 cm.

RUN — DON'T WALK from... the BIG BLOB...

...no, don't do that really. You need to get friendly with your big blob, and always use him for finding the perimeter of complicated shapes. He should always be willing to help — if not, just feed him a calculator or two and he'll be happy in no time. He's also quite partial to Camembert.

1) Turn over and write down what you have learnt.
2) Find the perimeter of the shape shown here.

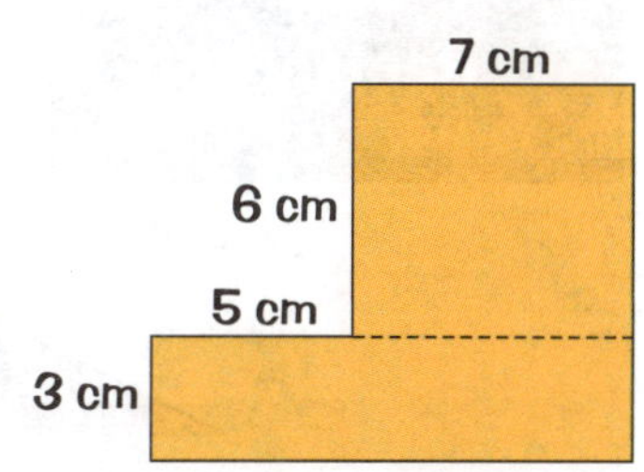

Areas

First things first — below are four basic area formulas. Best learn these unless you're planning on tipping bucketloads of easy exam marks down the toilet. (If you ARE planning on doing this, you'll need to raise your hand during the exam and ask to go to the bathroom — if they don't disappear on the first flush, pause, and then try again).

You must LEARN these Formulas:

1) RECTANGLE

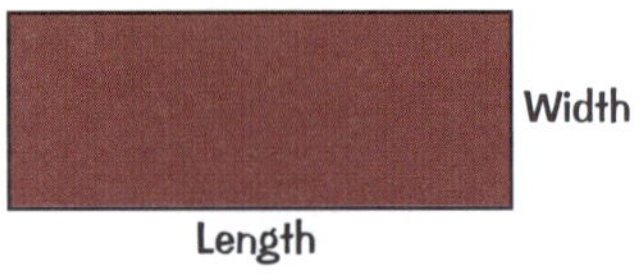

Area of RECTANGLE = length × width

$$A = l \times w$$

2) TRIANGLE

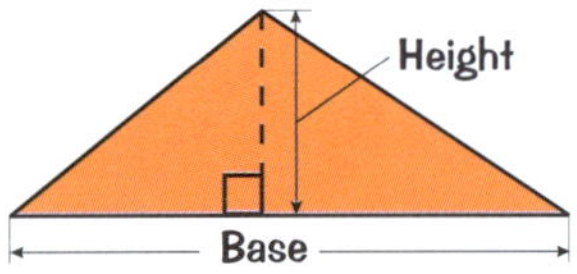

Area of TRIANGLE = ½ × base × vertical height

$$A = \frac{1}{2} \times b \times h_v$$

Note that the height must always be the vertical height, not the sloping height.

3) PARALLELOGRAM

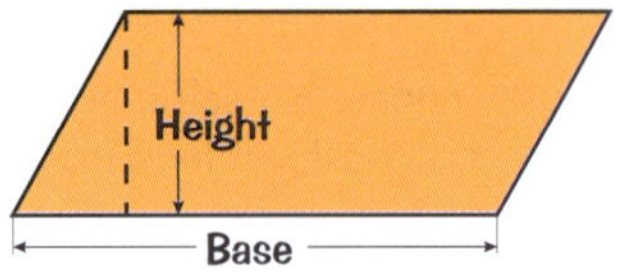

Area of PARALLELOGRAM = base × vertical height

$$A = b \times h_v$$

4) TRAPEZIUM

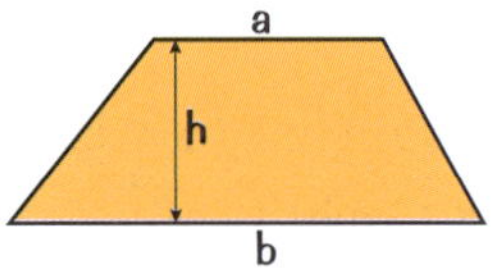

Area of TRAPEZIUM = average of parallel sides × distance between them

$$A = \frac{1}{2} \times (a + b) \times h$$

Areas of More Complicated Shapes

You often have to find the area of strange looking shapes in exam questions. What you always find with these questions is that you can break the shape up into simpler ones that you can deal with.

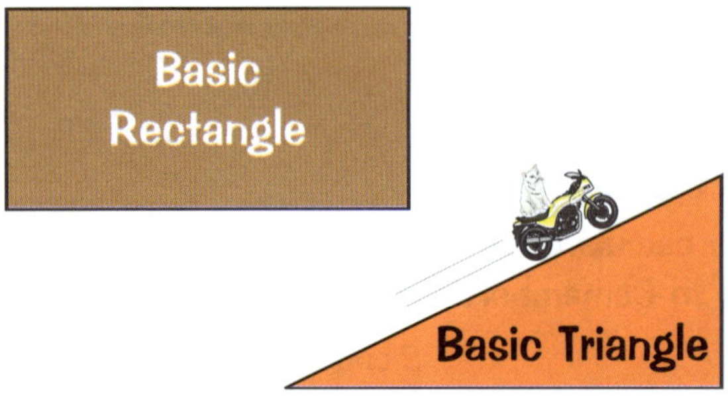

1) SPLIT THEM UP into the two basic shapes: RECTANGLE and TRIANGLE
2) Work out the area of each bit SEPARATELY
3) Then ADD THEM ALL TOGETHER

See next page for a lovely example...

Areas

...and here's the lovely example, as promised...

EXAMPLE: Work out the area of the shape shown:

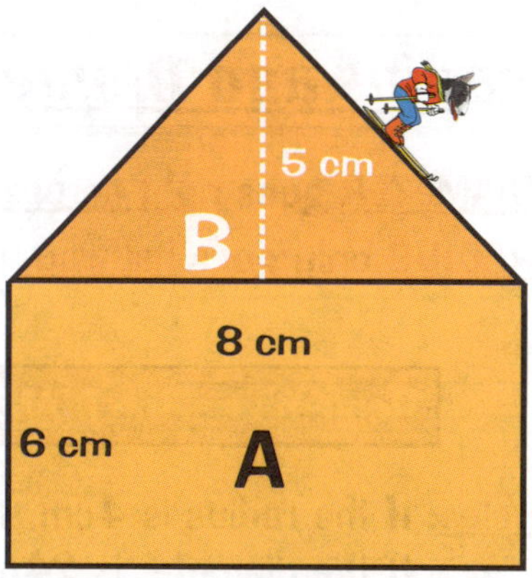

ANSWER:

Shape A is a rectangle:

$\text{Area} = L \times W$
$= 8 \times 6$
$= 48\ \text{cm}^2$

Shape B is a triangle:

$\text{Area} = \frac{1}{2} \times b \times h$
$= \frac{1}{2} \times 8 \times 5$
$= 20\ \text{cm}^2$

TOTAL AREA $= 48 + 20 = 68\ \text{cm}^2$

Don't Reach Straight for the Calculator

You might be kidding yourself that it 'takes too long' to write down your working out — but what's so great about getting ZERO MARKS for an easy question?

Compare these two answers for finding the area of the triangle opposite:

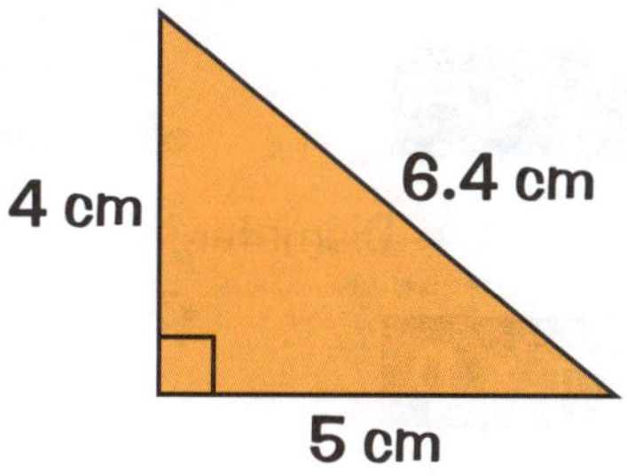

ANSWER 1: 20 ✗

ANSWER 1 gets NO MARKS AT ALL — 20 is the wrong answer and there's nothing else to give any marks for.

ANSWER 2: $A = \frac{1}{2} \times B \times H$ ✓
$= \frac{1}{2} \times 5 \times 4$ ✓
$= 10\ \text{cm}^2$ ✓

ANSWER 2 has 3 bits that all get marks, — so even if the answer was wrong it would still get most of the marks!

The thing is though, when you write it down step by step, you can see what you're doing and you won't get it wrong in the first place — try it next time, go on... just for the wild experience.

I need to take my shape to the barbers — it's getting area and area...

Hahaha.... Not amused? OK suit yourself. It's time to memorise the area formulas and learn how to deal with complicated shapes. Then find the areas of these 4 shapes...

1)

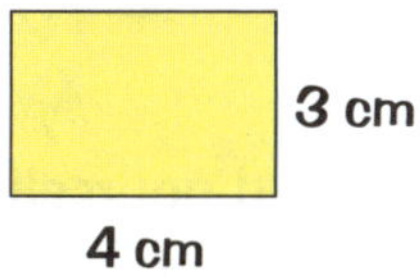

2)

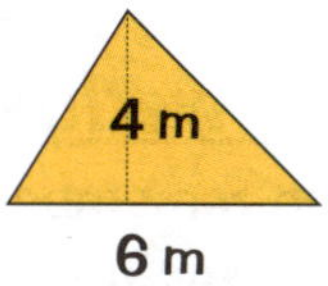

3)

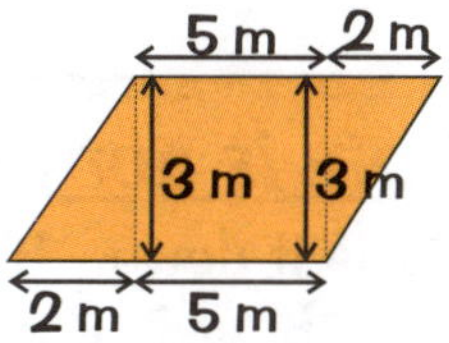

4)

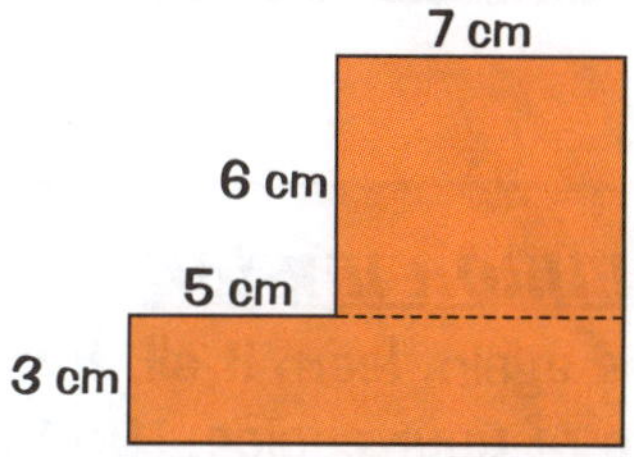

Circles

There's a surprising number of circle terms you need to know — don't mix them up. Oh, and it's probably best to have a snack before starting this page. All the talk of pi can make you a bit peckish.

1) Radius and Diameter

The DIAMETER goes right across the circle.

The RADIUS only goes halfway across.

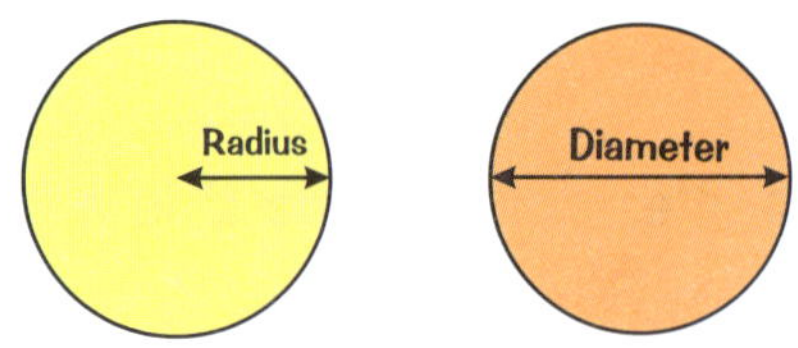

Remember: the DIAMETER IS EXACTLY DOUBLE THE RADIUS

Examples: If the radius is 4 cm, the diameter is 8 cm,
If the diameter is 24 m, the radius is 12 m.

2) Area, Circumference and π

The CIRCUMFERENCE is the distance round the outside of the circle.
You should know the formulas for area and circumference...

AREA of CIRCLE $= \pi \times (\text{radius})^2$

$$A = \pi \times r^2$$

e.g. if the radius is 4 cm, then
$A = 3.14 \times (4 \times 4)$
$= 50.24\text{ cm}^2$

CIRCUMFERENCE $= \pi \times$ Diameter

$$C = \pi \times D$$

$\pi = 3.141592....$
$= 3.14$ (approx)

The big thing to remember is that π (called "pi") is just an ordinary number (3.14159...) which is usually rounded off to 3.14.

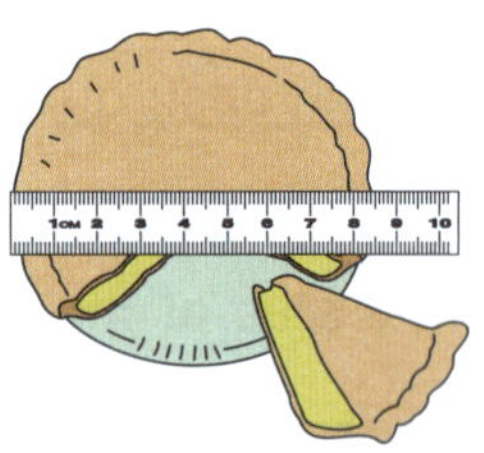

3) Tangents, Chords, Arcs and the Rest...

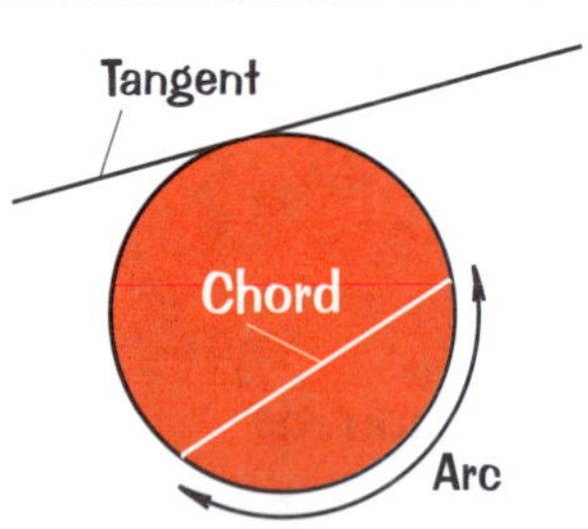

A TANGENT is a straight line that just touches the outside of the circle.

A CHORD is a line drawn across the inside of a circle.

AN ARC is just part of the circumference of the circle.

A SECTOR is a WEDGE SHAPED AREA (like a piece of cake) cut right from the centre.

SEGMENTS are the areas you get when you cut the circle with a chord.

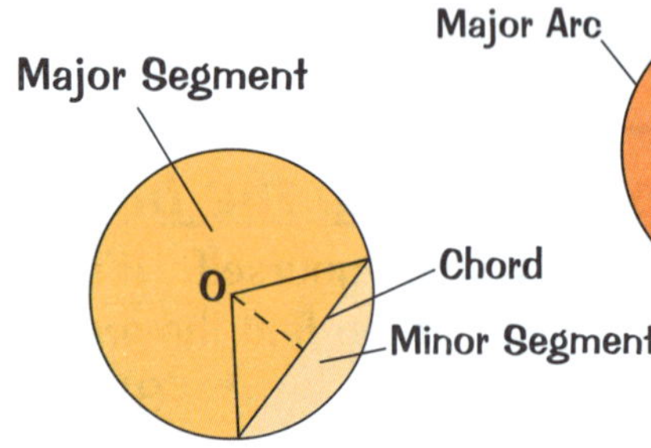

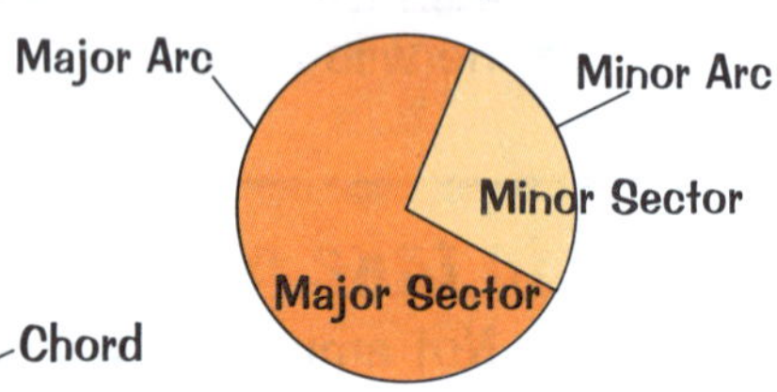

Common mistake — a slice of pie is not called a wedgie...

Once again, learn it all, turn over and scribble it down. If you can remember it all, reward yourself with a pie of your choice. Do I want a wedgie? Oh, well, if you're offering it's rude not to...

Circle Questions

There's a pretty big decision to make on this page — you might want to sit down...

The Big Decision: "Which circle formula do I use?"

1) If the question asks for 'the area of the circle', YOU MUST use the FORMULA FOR AREA:

$$A = \pi \times r^2$$

2) If the question asks for "circumference' (the distance around the circle) YOU MUST use the FORMULA FOR CIRCUMFERENCE:

$$C = \pi \times D$$

And remember, it makes no difference at all whether the question gives you the radius or the diameter, because whichever one they give you, it's DEAD EASY to work out the other one — the diameter is always DOUBLE the radius.

Example 1:

"Find the circumference and the area of the circle shown below."

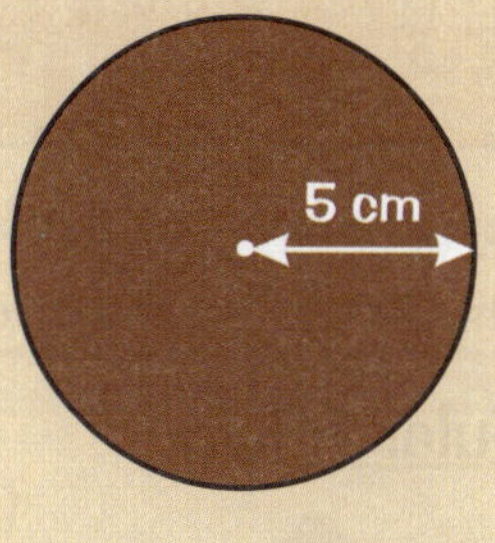

ANSWER:

The radius = 5 cm, so the Diameter = 10 cm (easy huh?)

Formula for circumference is:

$C = \pi \times D$, so

$C = 3.14 \times 10$

$= 31.4$ cm

Formula for AREA is:

$A = \pi \times r^2$

$= 3.14 \times (5 \times 5)$

$= 3.14 \times 25 = 78.5\text{ cm}^2$

Example 2: The good old "Wagon Wheel" question:

This is a very common Exam question.

"How many turns must a wheel of diameter 1.2 m make to go a distance of 12 m?"

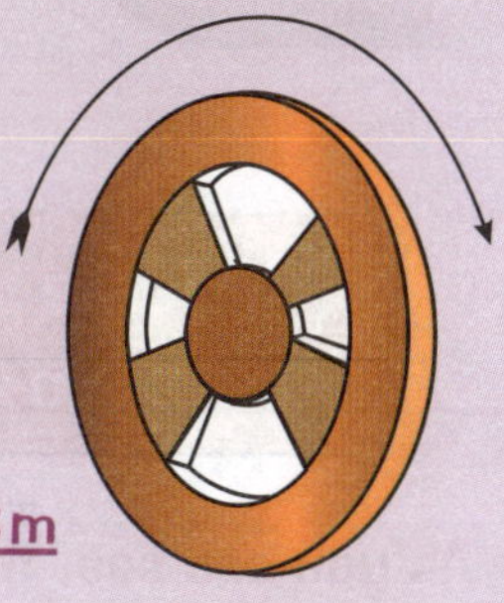

ANSWER:

Each full turn moves it one full circumference across the ground, so

1) find the circumference using "$C = \pi \times D$" : $C = 3.14 \times 1.2 = 3.768$ m

2) then find how many times it fits into the distance travelled, by dividing:

i.e. $12\text{ m} \div 3.768\text{ m} = 3.18$ so the answer is 3.2 turns of the wheel

History lesson — Wagon Wheels were a popular lunch box snack in 1993...

Learn this page so that you won't turn into a nervous wreck if you need to make 'The Big Decision' in the exam. And make sure you understand both the examples — something similar could come up.

1) A plate has a diameter of 14 cm. Find the area and circumference of it using the methods you have just learnt. Remember to show all your working out.

2) How many turns will a hoop of diameter 2 m make if it rolls 240 m?

Volume

I'm not going to say anything apart from...

VOLUMES — YOU MUST LEARN THESE TOO

1) Cuboid (rectangular block)

(This is also known as a 'rectangular prism' — see below to understand why.)

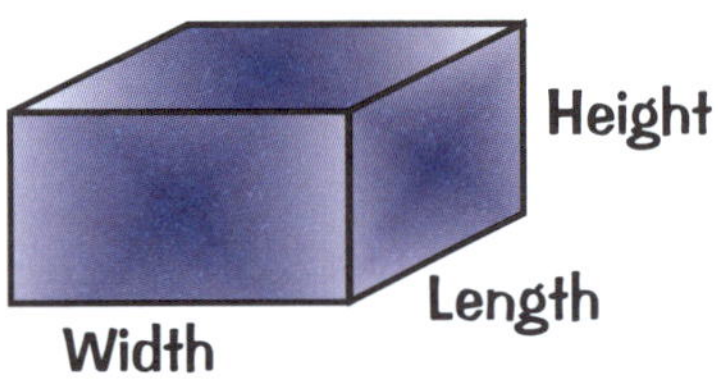

Volume of Cuboid = length × width × height

$$V = l \times w \times h$$

(The other word for volume is CAPACITY)

2) Prism

A PRISM is a solid (3-D) object which is the same shape all the way through — i.e. it has a CONSTANT AREA OF CROSS-SECTION.

Now, for some reason, not a lot of people know what a prism is, but they come up all the time in Exams, so make sure YOU know.

Triangular Prism

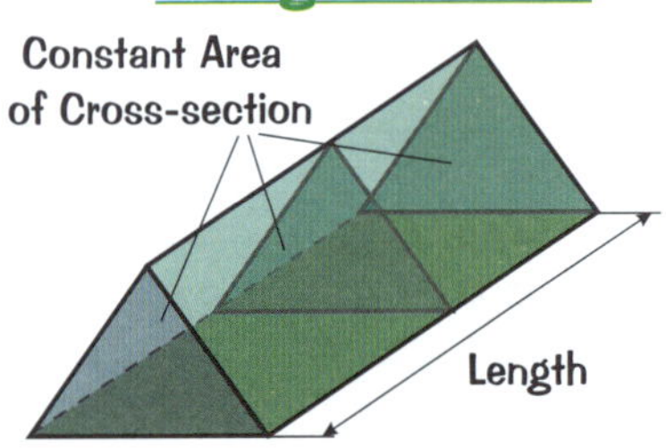

Hexagonal Prism

(a flat one, certainly, but still a prism)

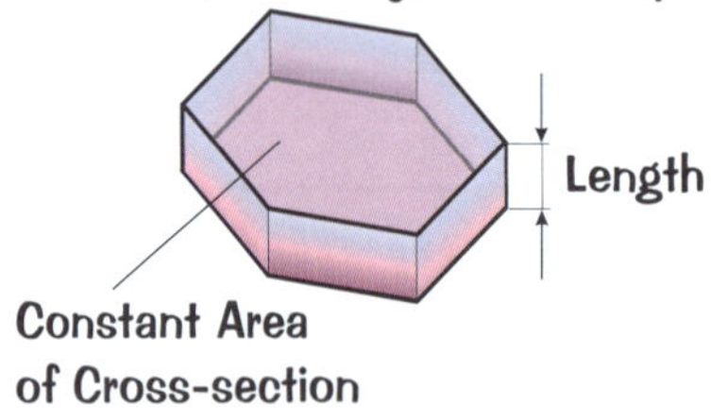

Circular Prism

(or Cylinder)

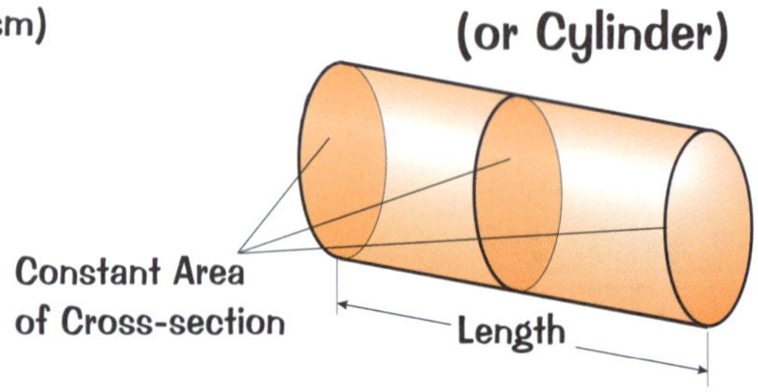

Volume of prism = Cross-sectional Area × length

$$V = A \times l$$

As you can see, the formula for the volume of a prism is very simple.
The difficult part, usually, is finding the area of the cross-section.

Don't make it any more angry — it's already a cross-section...

With things like this, once you've learnt the formula it's a doddle. You then just need to put the right numbers in and the answer is yours. All yours. Mwah-ha-ha-ha-haaaaaa.

Find the volume of each of these prisms:

a)

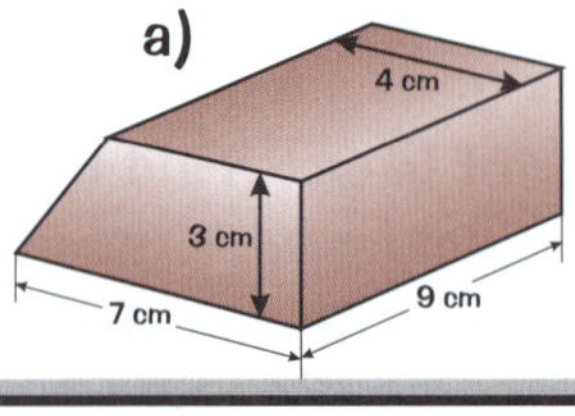

b)

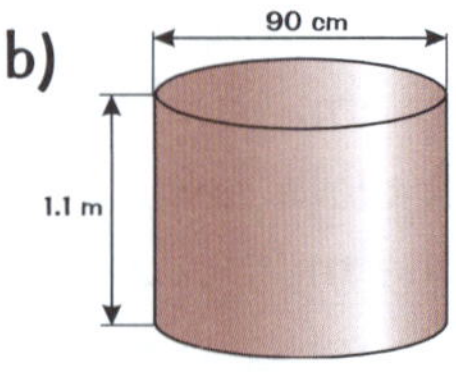

Solids and Nets

If you want to know what face, edge and vertex mean, you've come to the right place to find out...

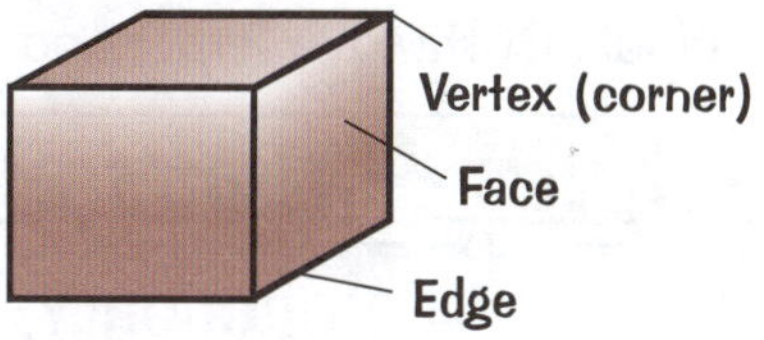

Surface Area and Nets

1) SURFACE AREA only applies to solid 3-D objects, and it's simply the total area of all the outer surfaces added together. If you were painting it, it's all the bits you'd paint.
2) There is never a simple formula for surface area — you have to work out each face in turn and then ADD THEM ALL TOGETHER.
3) A NET is just A SOLID SHAPE FOLDED OUT FLAT.
4) So obviously: SURFACE AREA OF SOLID = AREA OF NET.

There are 4 nets of common nets shown below, make sure you can find the area of each net.

1) Triangular Prism

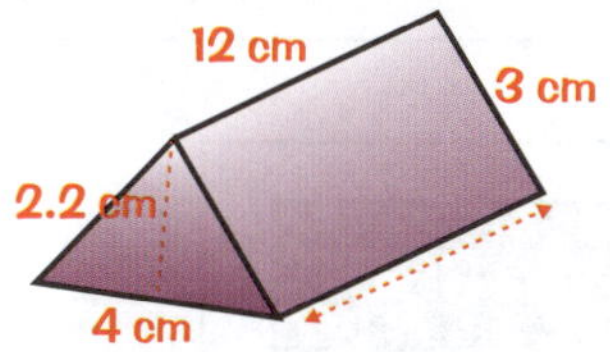

Triangular Prism

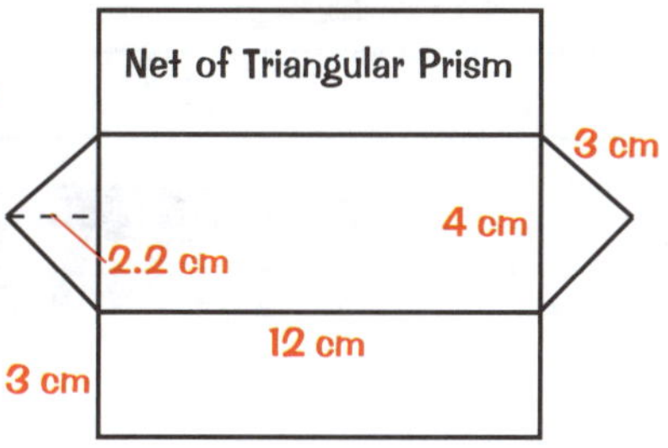

2) Cube

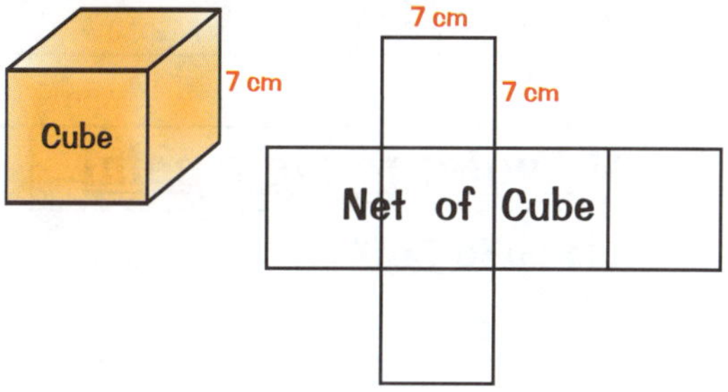

3) Cylinder

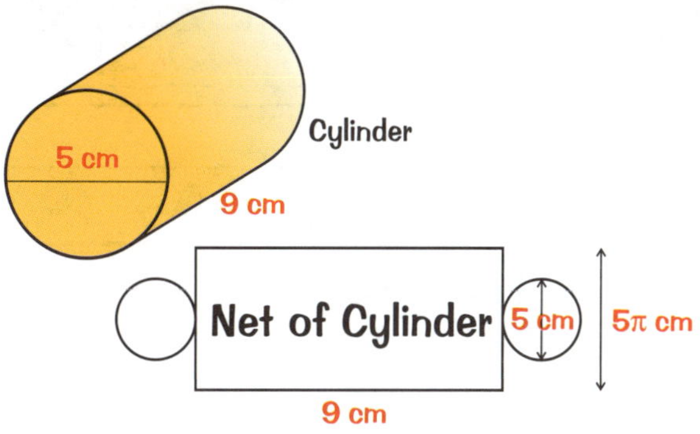

4) Pyramid

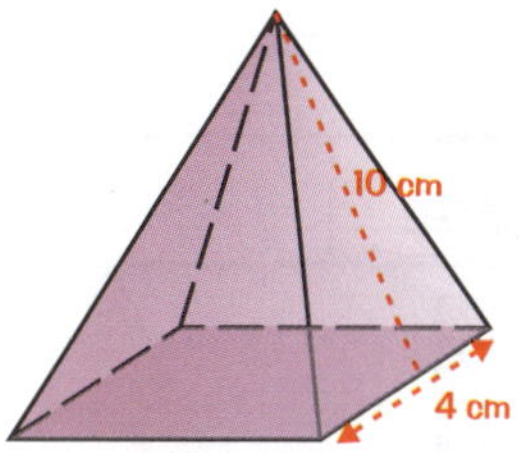

Square-based Pyramid

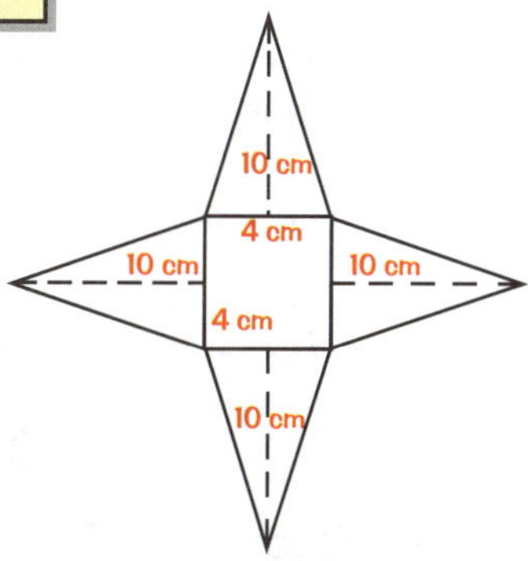

Net of Square-based Pyramid

Edge and vertex, face and toes, face and toes. Edge and vertex....

...face and toes, face and toes. And eyes and ears and mouth and nose. Edge and vertex, face and toes. Face and toes. I didn't make that version up — my nursery was just lame.

1) Work out the area of all four nets shown above.

Six Angle Rules

Rules, rules, rules. Gotta love 'em. And even if you don't, you gotta learn 'em. Try these for size:

6 Simple Rules — that's all:

If you know them ALL — THOROUGHLY, you at least have a fighting chance of working out problems with lines and angles. If you don't — you've no chance.

1) Angles in a triangle

Add up to 180°.

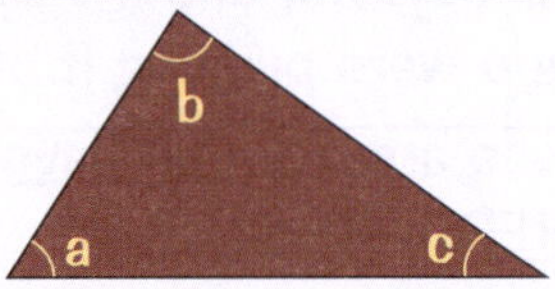

$a+b+c=180°$

2) Angles on a straight line

Add up to 180°.

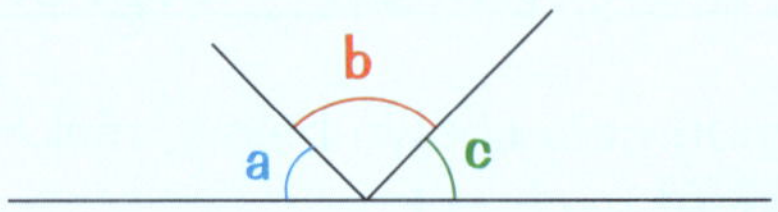

$a+b+c=180°$

3) Angles in a 4-sided shape

(a 'Quadrilateral')
Add up to 360°.

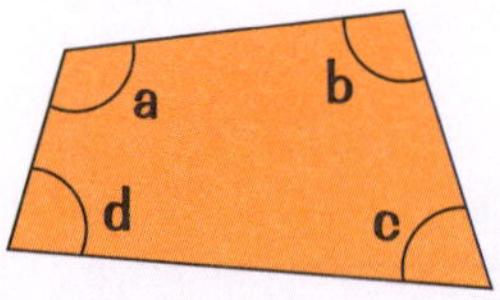

$a+b+c+d=360°$

4) Angles round a point

Add up to 360°.

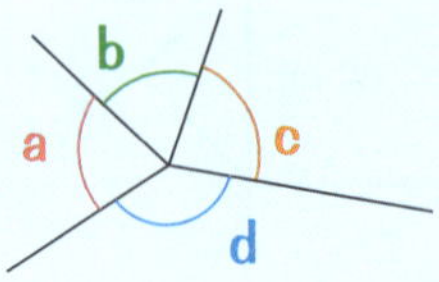

$a+b+c+d=360°$

5) Exterior Angle of Triangle

Exterior Angle of triangle = sum of Opposite Interior angles

i.e. $a+b=d$

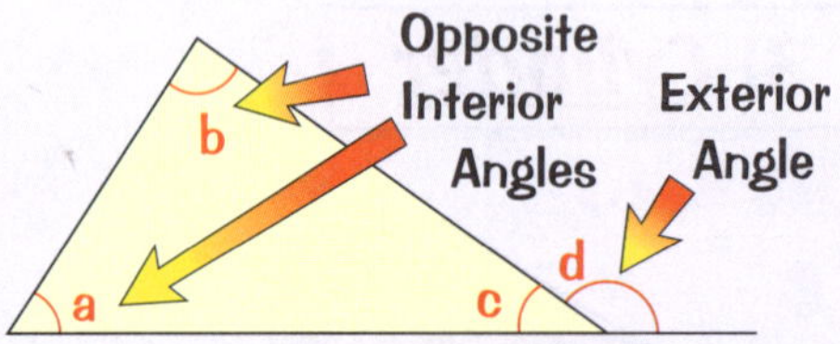

6) Isosceles triangles

2 sides the same
2 angles the same

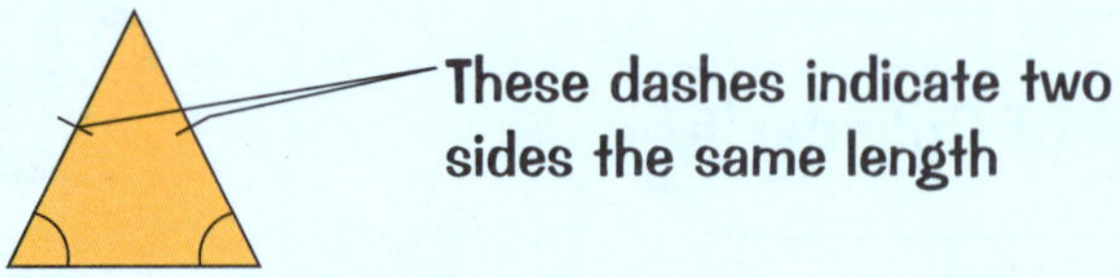

In an isosceles triangle, YOU ONLY NEED TO KNOW ONE ANGLE to be able to find the other two, which is very useful IF YOU REMEMBER IT.

a)

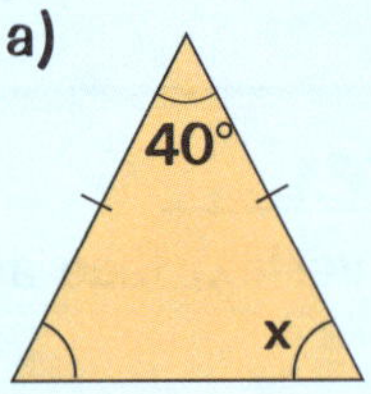

180° – 40° = 140°
The two bottom angles are both the same and they must add up to 140°, so each one must be half of 140° (= 70°). So x = 70°.

b)

The two bottom angles must be the same, so 50° + 50° = 100°.
All the angles add up to 180°
so y = 180° – 100° = 80°.

Parallel Lines

Up next — parallel lines. These make working out angles even easier. No really, they do...

Angles and Parallel Lines

Any two lines that cross will have two pairs of vertically opposite angles which are equal, like this:

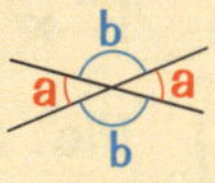

Whenever one line goes across 2 parallel lines, then the two bunches of angles are the same, as shown below:

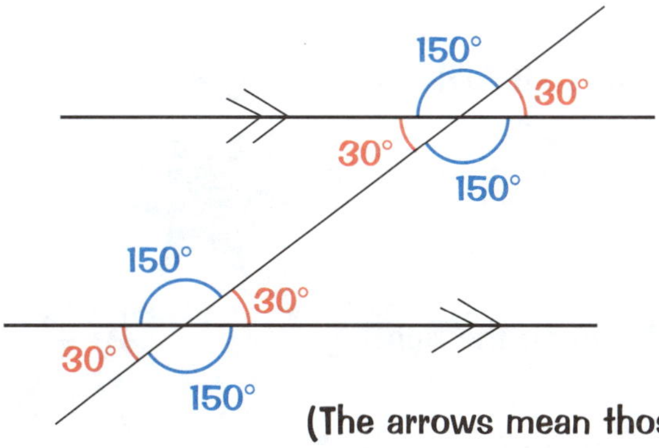

(The arrows mean those 2 lines are parallel)

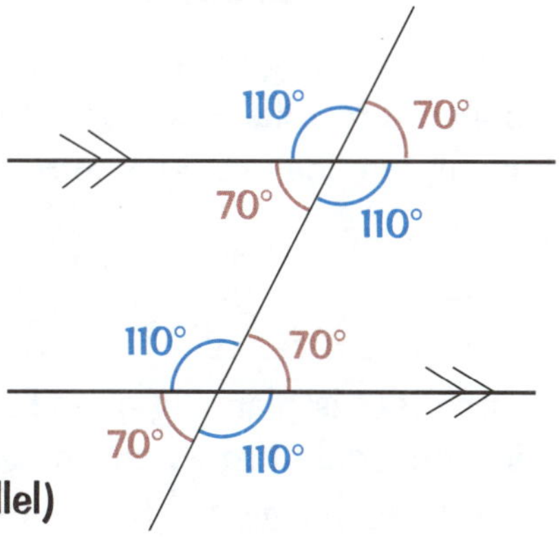

Whenever you have TWO PARALLEL LINES...
1) there are only two different angles: A SMALL ONE and A BIG ONE
2) and they ALWAYS ADD UP TO 180°. E.g. 30° and 150° or 70° and 110°

The trickiest bit about parallel lines is SPOTTING THEM IN THE FIRST PLACE — watch out for these 'Z', 'C', 'U' and 'F' shapes popping up. They're a dead giveaway that you've got a pair of parallel lines.

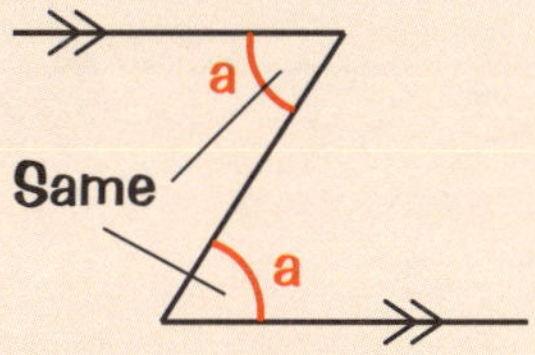

In a Z-shape they're called "ALTERNATE ANGLES"

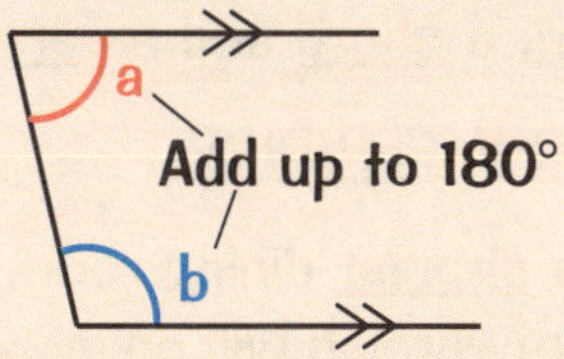

If they add up to 180° they're called "SUPPLEMENTARY ANGLES"

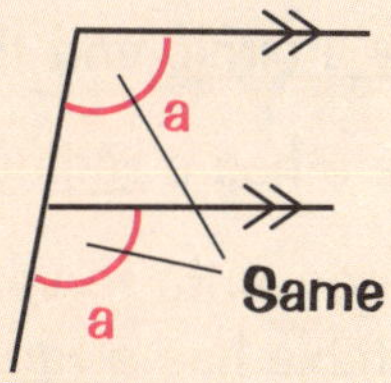

In an F-shape they're called "CORRESPONDING ANGLES"

Alas you're expected to learn these three silly names too!
Don't say 'F-angles' in the exam, say 'corresponding'...

I chose not to choose life. I chose parallel line spotting...

Spotting parallel lines is really, really important, so always have a good look for them.
There are some questions that you won't be able to answer if you miss them.
And I think we'd agree that that's a bad thing.

1) The diagram shown here has one angle given as 60°. Find all the other 7 angles.

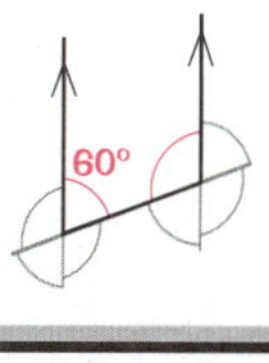

Revision Test for Unit 3 — 1

WHAT YOU'RE SUPPOSED TO DO HERE is put all the methods of this Unit so far into practice to answer these questions.

1) If $x^2 = 30$, find x accurate to one decimal place. (Trial and Improvement)

2) Plot the graph of $y = x^2 + 4x + 4$.

3) What are the two types of symmetry?

4) Sharon is making some sample birthday cards for a stationery company. She wants to fold each card along a line of symmetry, so it folds exactly together. Draw all the lines of symmetry for each of the card designs.
a) b) c) d)

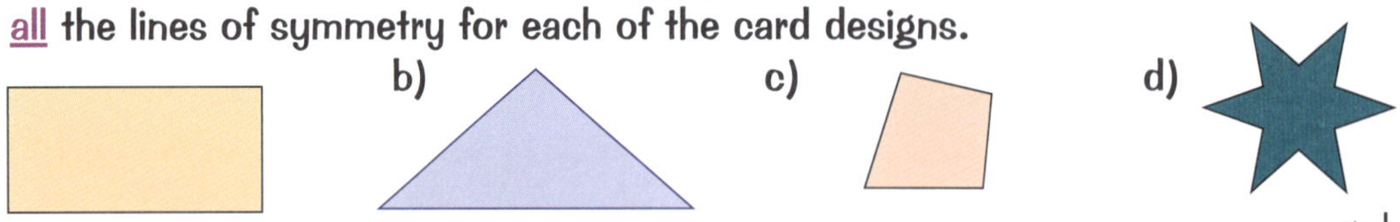

5) An architect is planning an extension to a house, shown on the right. Using the scale 1 m : 1 cm, draw:
a) the side elevation b) the front elevation c) the plan view.

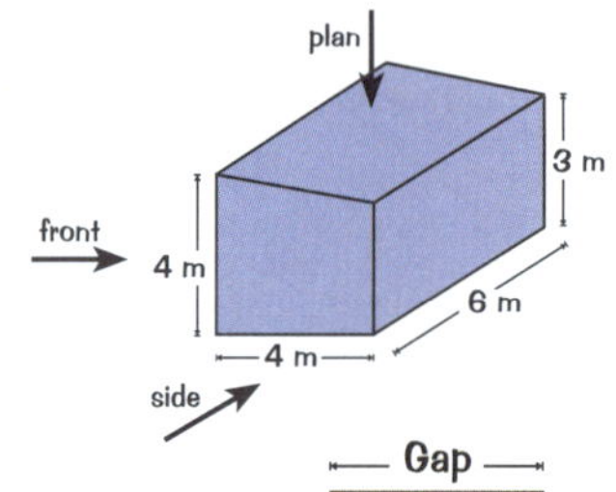

6) Work out the exterior and interior angles for a regular octagon (8 sides).

7) A plan of Alison's garden is shown on the right. Alison wants to put a fence around the perimeter of her garden, leaving only the gap that is shown. How many metres of fencing should Alison buy?

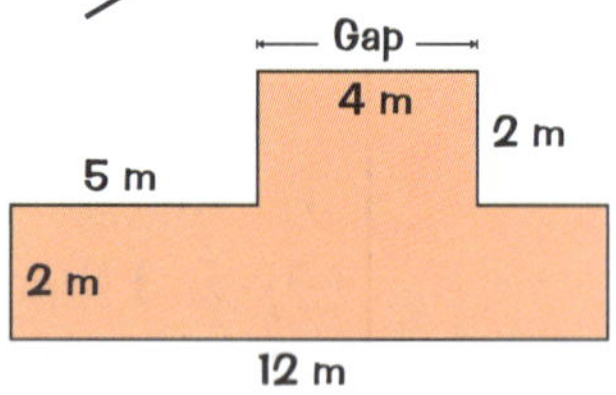

8) Clive is re-carpeting his lounge, which is rectangular. The room is 12 m long and 7 m wide. The carpet he wants costs £12 per m^2. How much will it cost Clive to carpet his lounge?

9) What are the TWO CIRCLE FORMULAS?

10) a) What is π ?
b) If a circle has a diameter of 12 m, what is its radius?

11) Draw a circle and then show on it a tangent, a chord and an arc.

12) A plate has a radius of 6 cm. Find its CIRCUMFERENCE.

13) Delene wants to varnish the top of her new circular dining room table. The table has a radius of 1.25 m. One tin of varnish will cover 2 m^2. How many tins of varnish must Delene buy?

14) Jane has been asked to design packaging to hold 30 cm^3 of washing powder. The packaging must be a prism. Sketch a possible design for the box, showing dimensions.

15) Do a sketch of these four solids and draw the net for each one:
a) A cube b) A cylinder c) A triangular prism d) A square-based pyramid

16) Work out angles X and Y in the diagram:

17) Give the correct names for 'Z', 'C', 'U' and 'F' angles and draw a sketch of each to explain what they are.

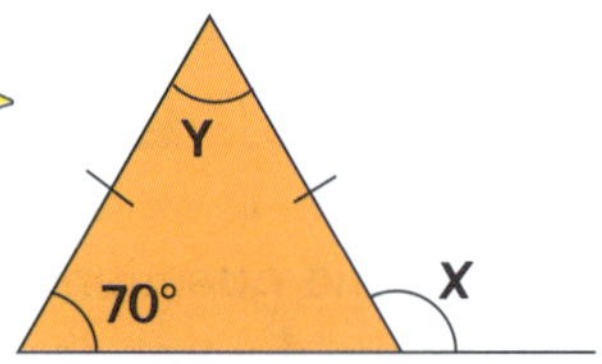

Congruence and Similarity

Shapes can be similar or congruent. And I bet you really want to know what that means — I can already picture your eager face. Well, lucky you — I've written a page all about it.

Congruent — Same Shape, Same Size

Congruence is another ridiculous maths word which sounds really complicated but it's not:

If two shapes are CONGRUENT, they are simply THE SAME — the SAME SIZE and the SAME SHAPE.

That's all it is. Just make sure you know the word.

CONGRUENT: Same size, same shape.
Note – you can have mirror images.

Similar — Same Shape, Different Size

Similar has a special meaning in maths, but nothing complicated:

If two shapes are SIMILAR they are exactly the SAME SHAPE but DIFFERENT SIZES.

SIMILAR: Same shape, different size.

Remember when you have similar shapes that the angles are always the same.

Two similar shapes can be seen as an 'enlargement'... see next page.

Pizza and a frisbee — same shape and size but don't get them confused...

Know what congruence and similar mean inside-out — it's important you don't get them mixed up.

1) a) Which of these four shapes are similar?
   b) Which are congruent?

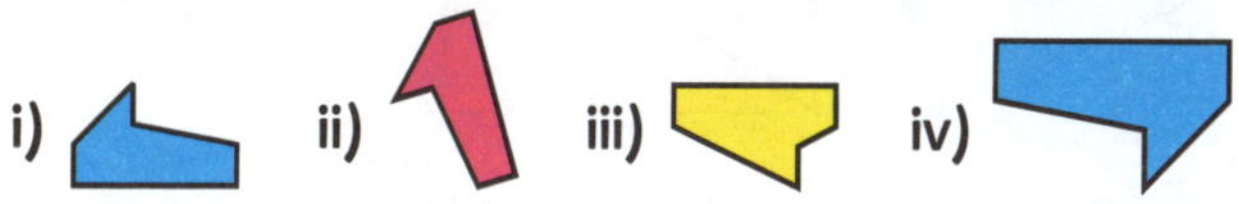

The Four Transformations

Objects can be transformed by changing their position or size, spinning or reflecting them.

Four Types of Transformation

Translation — ONE Detail
Enlargement — TWO Details
Rotation — THREE Details
Reflection — ONE Detail
Y (The Y doesn't stand for anything)

1) Use the word TERRY to remember the 4 types.
2) You must always give all the details for each type.

1) Translation

A translation is just a SLIDE. You must specify how far along and how far up the translation is.

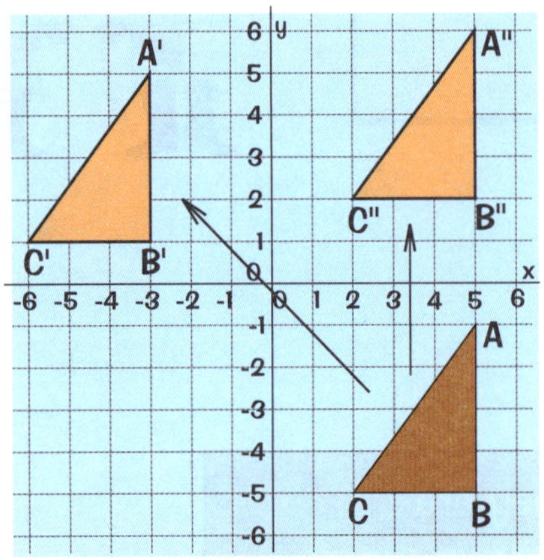

ABC to A'B'C' is a translation of 8 left and 6 up.
ABC to A''B''C'' is a translation of 7 up.

You might be asked to describe translations with vectors which look like this. $\begin{pmatrix} x \rightarrow \\ \uparrow y \end{pmatrix}$
x is the number of spaces right,
y is the number of spaces up.

As vectors, the translations shown in the diagram are: $\begin{pmatrix} -8 \\ 6 \end{pmatrix}$ and $\begin{pmatrix} 0 \\ 7 \end{pmatrix}$

2) Enlargement

You must give these 2 details: 1) The SCALE FACTOR
2) The CENTRE of Enlargement

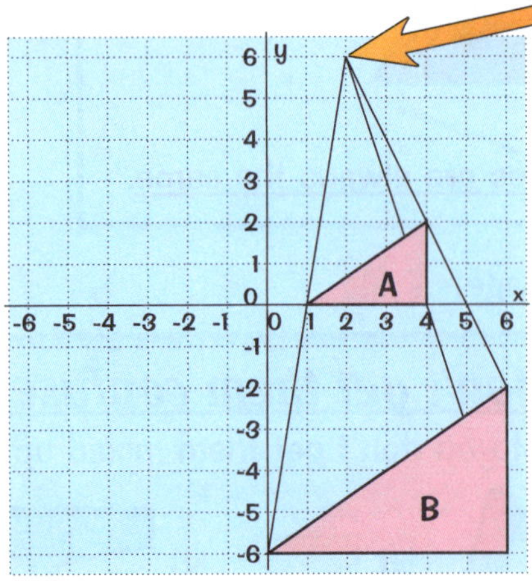

From A to B is an enlargement of scale factor 2, and centre (2,6)

N.B. Lengths doubled, distances from centre doubled too.

With enlargement, the ANGLES of the object remain unchanged. The RATIOS of the lengths of the sides, and the object's ORIENTATION remain unchanged. The size and position do change.

The Four Transformations

Two down, two to go...

3) Rotation

You must give these 3 details:
1) ANGLE turned
2) DIRECTION (Clockwise or Anti-clockwise)
3) CENTRE of Rotation

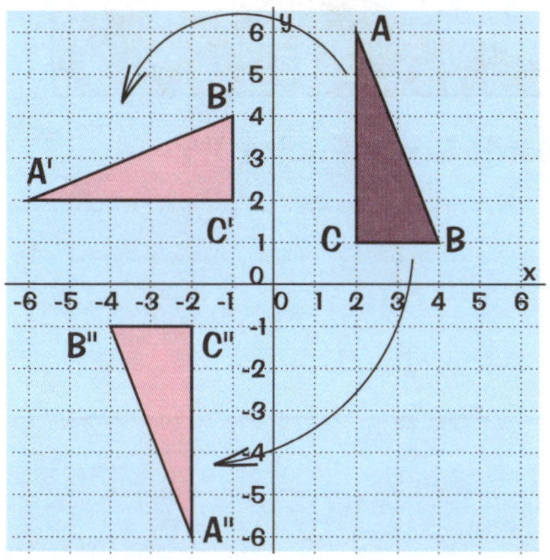

ABC to A'B'C' is a Rotation of 90°, anticlockwise, ABOUT the origin.

ABC to A''B''C'' is a Rotation of half a turn (180°), clockwise, ABOUT the origin.

(For half-turns, it doesn't actually matter if you go clockwise or anticlockwise.)

The only things that change in a rotation are the POSITION and the ORIENTATION of the object. Everything else remains unchanged.

4) Reflection

You must give this ONE detail: The MIRROR LINE

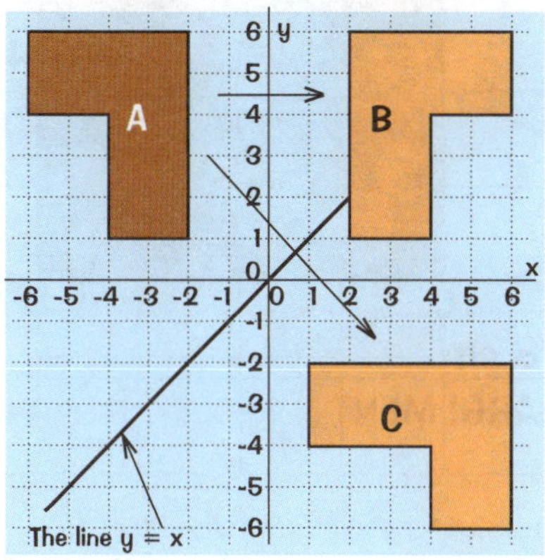

A to B is a reflection IN the y-axis.

A to C is a reflection IN the line $y=x$.

With reflection, the POSITION and ORIENTATION of the object are the only things that change.

Remember — you can ask for tracing paper in your exam to help you with transformations.

Wish someone would translate these pages for me...

Another two pages down. Hurrah. Before you move on you need to:
LEARN the names of the Four Transformations and the details for each.
Then, when you think you know it, turn over and write it all down.
Finally, do a dance to celebrate. Any dance will do.

1) Describe fully these 4 transformations:

A→B, B→C, C→A, A→D

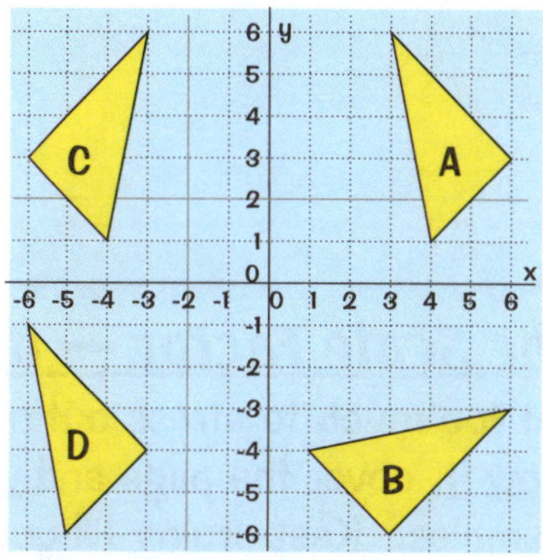

Enlargements

Enlargements are a really important type of transformation — so much so that here's an extra two pages on them, created especially for you. I know, I know, I'm too kind...

Scale Factors

If the Scale Factor is BIGGER THAN 1 then the shape gets BIGGER.

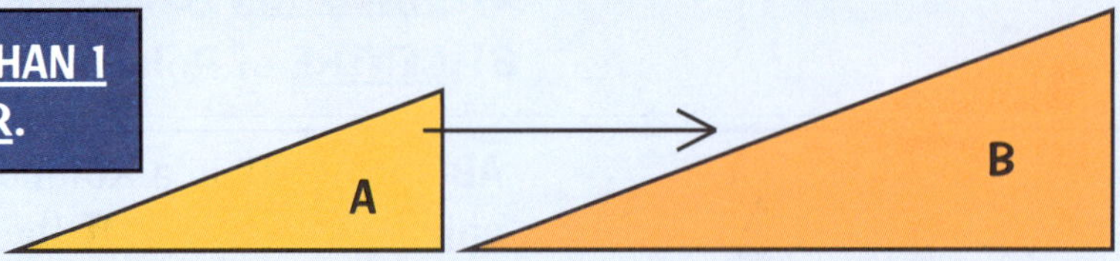

A to B is an Enlargement, Scale Factor 1½

The Scale Factor also tells you the RELATIVE DISTANCE of old points and new points from the Centre of Enlargement.

This is very useful for drawing an enlargement — you can use it to trace out the positions of the new points from the centre of enlargement, as shown in the diagram.

So for an enlargement on a grid, like the one here:

1) Work out how far away each of the corners is from the centre of enlargement
2) Multiply by the scale factor
3) Then draw in the new corners at this distance away from the centre.

Enlargement Scale Factor 4

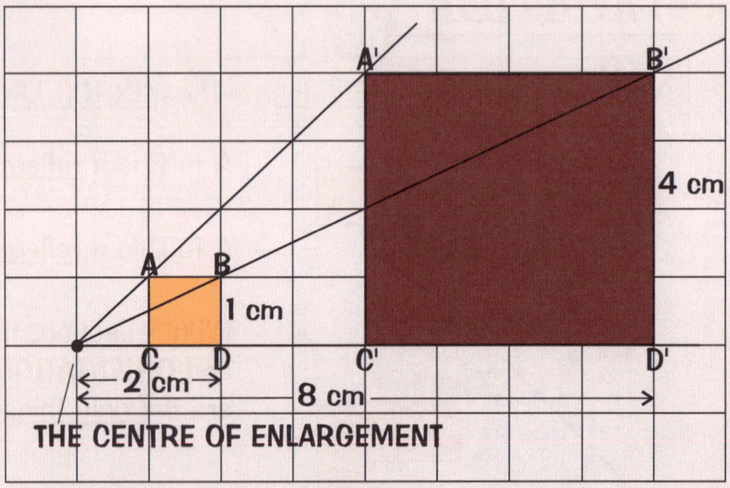

E.g. Corner B is 2 cm along and 1 cm up from the centre.
Corner B' is 8 cm along and 4 cm up from the centre.

The Scale Factor — it's time... to face... the diet...

Not too much to stress out over here — LEARN everything on the page, then when you think you know it, cover the page and write it all down from memory. Make sure you include the sketches and examples. Keep trying till you can. What a fun game that'll be. I'm almost jealous. Almost...

Enlargements

More on enlargements...

Use the Formula Triangle for Calculations

The lengths of the big and small shapes are related to the Scale Factor by this very important Formula Triangle which you must learn:

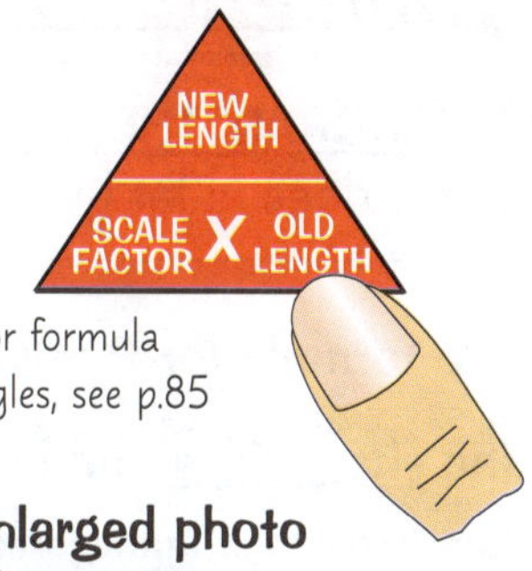

For formula triangles, see p.85

EXAMPLE: Find the missing width, x, in the diagram.

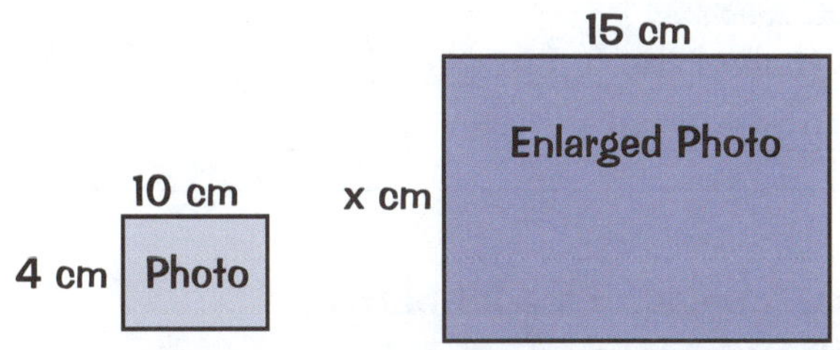

To find the width of the enlarged photo we use the formula triangle TWICE, (firstly to find the Scale Factor, and then to find the missing side):

1) Scale Factor = New length ÷ Old length = 15 ÷ 10 = 1.5
2) New width = Scale Factor × Old width = 1.5 × 4 = 6 cm

But without the formula triangle you're scuppered (Just like this fish)

You can also use the formula triangle to find perimeters in exactly the same way as lengths.
E.g. a square with sides 1 cm is enlarged by a scale factor of 2. What is the perimeter of the new shape?

Old perimeter = 1 × 4 = 4 cm
New perimeter = scale factor × old perimeter = 2 × 4 = 8 cm.

Areas and Volumes of Enlargements

Ho ho! This little joker catches everybody out.
The increase in area and volume is BIGGER than the scale factor.
For example, this box has been enlarged by Scale Factor 2.
You can see that the lengths are twice as big, each area is 4 times as big, and the volume is 8 times as big. The rule is this:

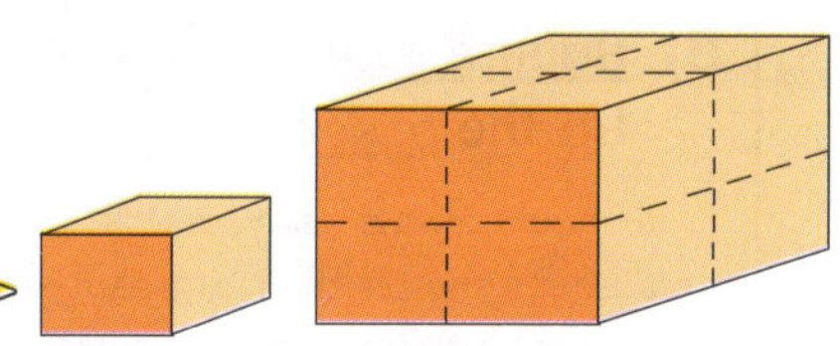

For a Scale Factor n:		e.g. 3:
The SIDES are	n times bigger	3 times bigger
The AREAS are	n^2 times bigger	3^2 = 9 times bigger
The VOLUMES are	n^3 times bigger	3^3 = 27 times bigger
	Simple... but VERY FORGETTABLE	

"Twice as much learning, 4 times better results, 8 times more fun...

...with a scale factor of 2, CGP's the guide for you". It's our new slogan. Pretty good isn't it...

1) Two triangles which are similar have heights of 5 cm and 45 cm respectively. The smaller triangle has an area of 30 cm². Find the area of the larger triangle.
2) Two similar cones have base diameters of 20 cm and 50 cm. If the volume of the smaller one is 120 cm³, find the volume of the other one.

Pythagoras' Theorem

Once upon a time there lived a clever chap called Pythagoras. He came up with a clever theorem...

Pythagoras' Theorem is Used on Right Angled Triangles

PYTHAGORAS' THEOREM is a handy little formula for RIGHT-ANGLED TRIANGLES.
What it does is let you find the length of the third side when you know two of them.

The formula for Pythagoras' theorem is: $a^2 + b^2 = h^2$ where a and b are the short sides and h is the long side of the triangle (called the hypotenuse)

Remember that Pythagoras can only be used on RIGHT-ANGLED TRIANGLES.

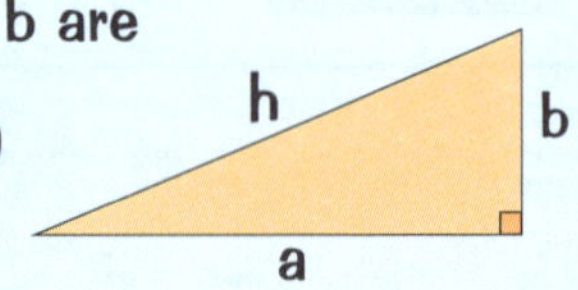

The trouble is, the formula can be quite difficult to use. Instead, it's a lot better to just remember these three simple steps, which work every time:

1) Square Them — SQUARE THE TWO NUMBERS that you are given, (use the x^2 button if you've got your calculator.)

2) Add or Subtract — To find the longest side, ADD the two squared numbers. To find a shorter side, SUBTRACT the smaller one from the larger.

3) Square Root — Once you've got your answer, take the SQUARE ROOT (use the √ button on your calculator).

EXAMPLE 1: "Find the missing side in the triangle shown."

1) Square them: $5^2 = 25$, $3^2 = 9$
2) You want to find a shorter side, so SUBTRACT: $25 - 9 = 16$
3) Square root: $\sqrt{16} = 4$
So the missing side = 4 m

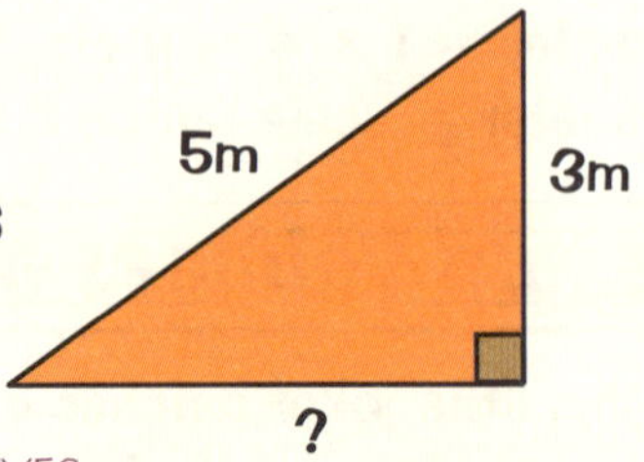

(You should always ask yourself: "Is it a sensible answer?" — in this case you can say "YES, because it's shorter than 5 m, as it should be since 5 m is the longest side, but not too much shorter")

The fish survived. Hurrah.

EXAMPLE 2: "Find the length of the line segment shown."

1) Work out how far across and up it is from A to B
2) Treat this exactly like a normal triangle...
3) Square them: $3^2 = 9$, $4^2 = 16$
4) You want to find the longest side (the hypotenuse), so ADD: $9 + 16 = 25$
5) Square root: $\sqrt{25} = 5$
So the length of the line segment = 5 units

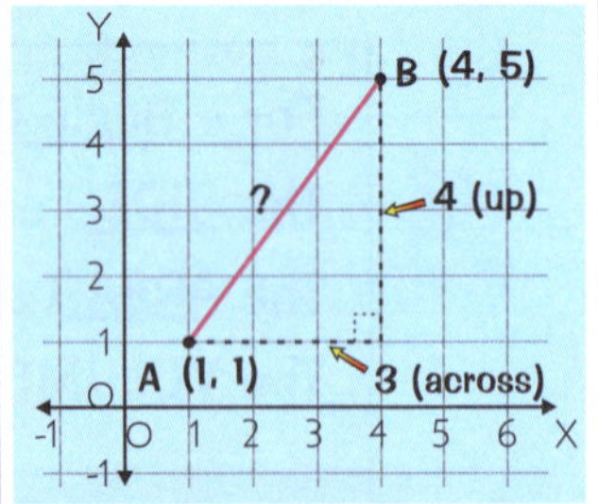

Hurrah, huzzah and all things joyful — 'tis the end of Pythagoras...

...apart from in the revision test later on, and in your exam, but don't think about that yet.

1) Apply the above method to find the missing side BC:
2) Another triangle has sides of 5 m, 12 m and 13 m.
Is it a right-angled triangle? How do you know?

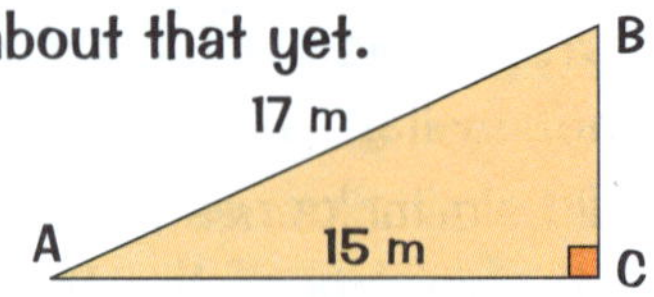

Formula Triangles and Speed

Formula triangles are extremely useful tools for lots of tricky maths problems.
They're very easy to use and very easy to remember. But don't just take my word for it...

If 3 things are related by a formula that looks either

like this: $A = B \times C$ or like this: $B = \frac{A}{C}$

then you can put them into a FORMULA TRIANGLE like this:

1) First decide where the letters go:

1) If there are TWO LETTERS MULTIPLIED TOGETHER in the formula then they must go ON THE BOTTOM of the Formula Triangle (and so the other one must go on the top). For example the formula 'F = m×a' fits into a formula triangle like this

2) If there's ONE THING DIVIDED BY ANOTHER in the formula then the one ON TOP OF THE DIVISION goes ON TOP IN THE FORMULA TRIANGLE (and so the other two must go on the bottom — it doesn't matter which way round).
So the formula 'Speed = Distance / Time' fits into a formula triangle like this.

2) Using the Formula Triangle:

Once you've got the formula triangle sorted out, the rest is easy:

1) COVER UP the thing you want to find and just WRITE DOWN what's left showing.
2) PUT IN THE VALUES for the other two things and just WORK IT OUT (look at the example below).

Speed = Distance ÷ Time

This probably comes up every single year, and they never give you the formula.
Luckily there's an easy FORMULA TRIANGLE to help you:

Of course you still have to remember the order of the letters in the triangle (S^DT) — but this time we have the word SoDiT to help you.

So if it's a question on speed, distance and time just say: **SOD IT**.

EXAMPLE: "A car travels 90 miles at 36 miles per hour. How long does it take?"

ANSWER: We want to find the TIME, so cover up T in the triangle which leaves D/S,

so T = D/S = Distance ÷ speed = 90÷36 = 2.5 hours

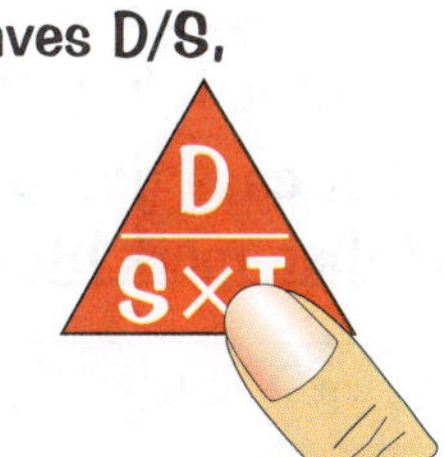

LEARN THE FORMULA TRIANGLE, AND YOU'LL FIND QUESTIONS ON SPEED, DISTANCE AND TIME VERY EASY.

Formula triangles — it's all a big cover up...

By now, you should be convinced that formula triangles are the most amazingly useful thing ever. They'll help make using formulas much easier. The best thing to do is learn the whole page, cover it up and scribble down everything you can remember. It's a really good way to help you learn everything.

Maps and Map Scales

If maps were the same size as the place they were showing, then they'd be big... and a bit pointless. Thankfully maps are usually made using scales. The most usual map scale is '1 cm = so many km'. This just tells you how many km in real life it is for 1 cm measured on the actual map itself.

1) Converting 'cm on the Map' into 'Real km'

This map shows the original Roman M6 Motorway built by the Emperor Hadrian in the year AD120.

The scale of the map is '1cm to 8km'
"Work out the length of the section of M6 between Wigan and Preston."

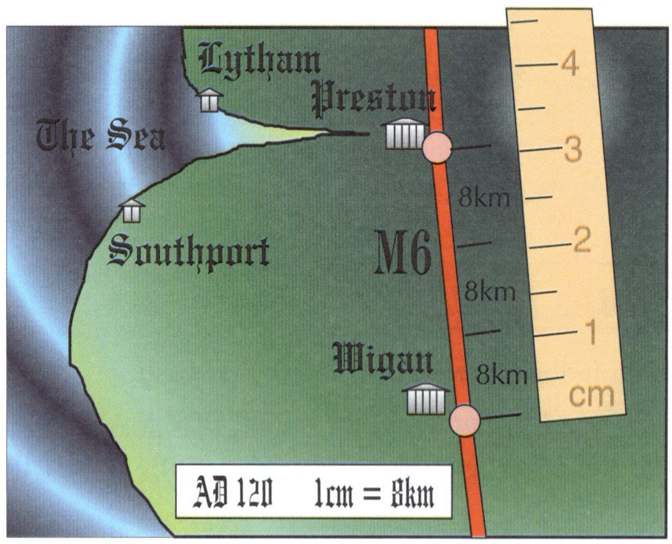

This is what you do (as shown on the diagram)

1) PUT YOUR RULER AGAINST THE THING you're finding the length of
2) MARK OFF EACH WHOLE CM AND WRITE IN THE DISTANCE IN KM next to each one
3) ADD UP ALL THE KM DISTANCES TO FIND THE WHOLE LENGTH of the road in km. (i.e. 8km + 8km + 8km = 24km)

Of course if they just tell you the thing is, say, 5 cm long you won't be able to put your ruler on it. For example, they could say "The distance between Wigan and Southport on the map is 5 cm. How far is it in real life?"

In that case you should draw a line 5 cm long and then mark off the km on it using your ruler just the same:

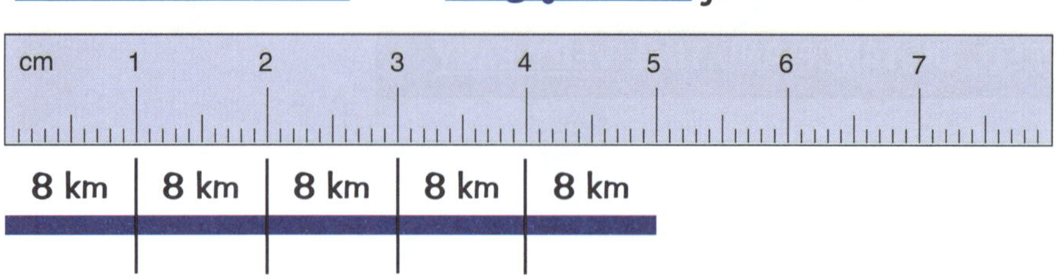

Picture of 80's mobile.
Scale: 1 cm = 3.6 km

So 5 cm on the map = 8 + 8 + 8 + 8 + 8 = 40 km.

Maps and Map Scales

2) Converting 'Real km' into 'cm on the Map'

Example:

"A map is drawn on a scale of 1cm to 2km.
If a road is 12km long in real life,
how long will it be in cm on the map?"

Answer:

1) Start by drawing the road as a straight line:

2) Mark off each cm and fill in how many km it is for each one

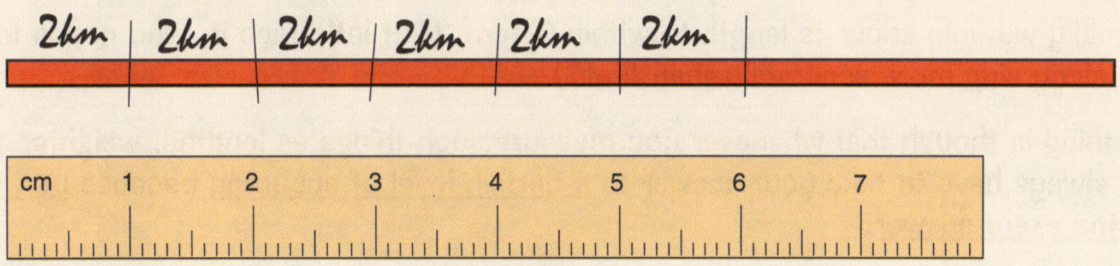

3) Keep going until the km add up to the full distance (12km in this case).

Then just count how many cm long your line is (In this case 6cm).

Instead of a map, you might get a question about a plan in your exam — like a plan of someone's bedroom or a garden. You do them in just the same way — the only difference is that the scale will probably be in m rather than km, e.g. 1cm to 2m. You might even be asked to draw something on a plan — just use the method above to work out how big it should be on the plan, then draw it.

What's a Roman soldier's favourite meal? Caesar salad...

If you're asked a map question in your exam — just remember the 3 rules for working with map scales. A ruler will be handy too. You're allowed to take them into exams. No need to hide one up your sleeve.

1) Work out the length in m of the runway shown here:

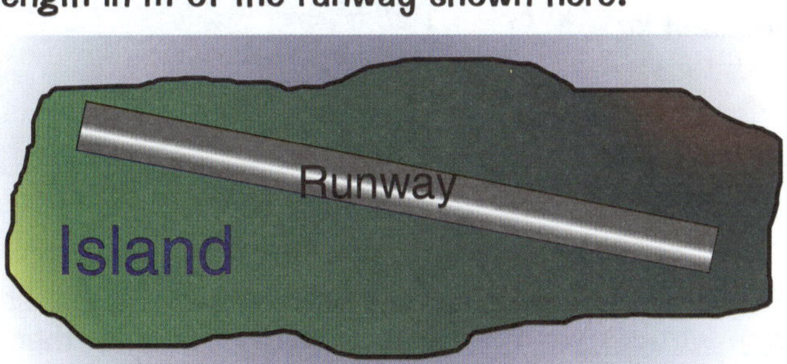

SCALE: 1cm to 200m

2) How many cm long would a 500m runway be?

Rounding Off Measurements

A lot of things you measure have a value which you can never know exactly, no matter how carefully you try and measure them.

Measurements Need to be Rounded Off

Take this slimy grey slug for example:

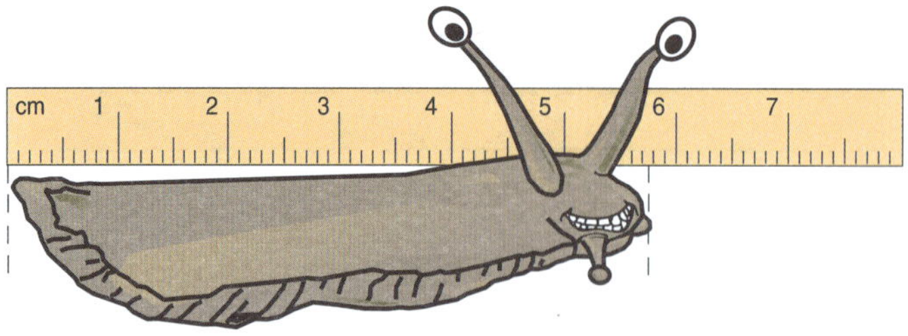

It has a length somewhere between 5 cm and 6 cm and if you look closer you can even say it's somewhere between 5.7 cm and 5.8 cm, but you can't really tell any more accurately than that.

So really we only know its length to within 0.1 cm. (But let's face it, who needs to know the length of a slimy slug more accurately than that?)

The thing is though that whenever you measure such things as lengths, weights, speeds etc, you always have to take your answer to a certain level of accuracy because you can never get the exact answer.

The simple rule is:

You always round off to the number that it's NEAREST TO

If we take our slimy slug, then to the nearest cm his length is 6 cm (rather than 5 cm) and to the nearest 0.1 cm it's 5.8 cm (rather than 5.7)

Postage Rates

All this about rounding off to the nearest number is all well and good, but in the Exam they can cheerfully spring a question on you about how much it costs to send a parcel and then the rules are completely different.

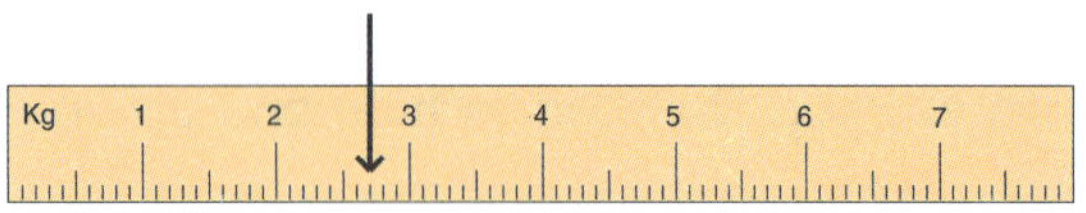

3 kg to the nearest kg

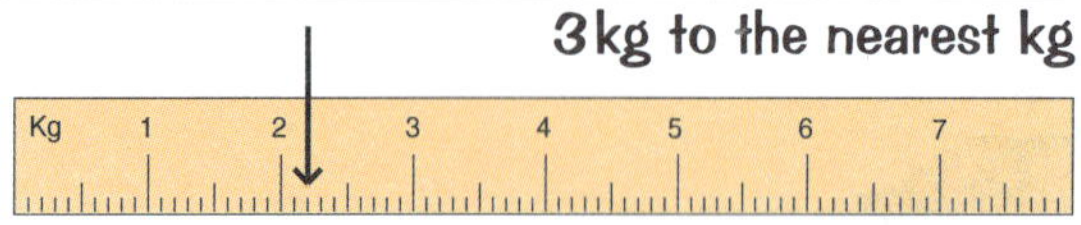

2 kg to the nearest kg

POSTAGE RATES

Weight not over:	Price
1 kg	£2.70
2 kg	£3.30
3 kg	£4.10

One reading would round off to 3 kg and the other to 2 kg but the postage rate is set by 'WEIGHT NOT OVER...' so both of these would cost the same (£4.10) because they are both more than 2 kg but less than 3 kg. It's tricky, so watch out for questions like that.

Conversion Factors

Conversion Factors are a really good way of dealing with all sorts of questions and the method is dead easy. Prepare yourself for the 3 steps towards conversion-related happiness...

Method

1) Find the CONVERSION FACTOR (always easy).
2) Multiply AND divide by it.
3) Choose the COMMON SENSE ANSWER.

Example 1:

"A Giant Sea-slug called Kevin was washed up near Grange-Over-Sands. He was 18.6 m in length. How long is this in cm?"

Step 1) Find the CONVERSION FACTOR
In this question the Conversion Factor = 100
— simply because 1 m = 100 cm

Step 2) MULTIPLY AND DIVIDE by the conversion factor:
18.6 m × 100 = 1860 cm (makes sense)
18.6 m ÷ 100 = 0.186 cm (ridiculous)

Step 3) Choose the COMMON SENSE answer:
Obviously the answer is that 18.6 m = 1860 cm

Check page 91 for the conversion factors.

Example 2:

"If £1 is equal to 1.7 US Dollars, how much is 63 US Dollars in £s?"

Step 1) Find the CONVERSION FACTOR
In this question the Conversion Factor is obviously 1.7
(When you're changing foreign money it's called the 'Exchange Rate')

Step 2) MULTIPLY AND DIVIDE by the conversion factor:
63 × 1.7 = 107.1 = £107.10
63 ÷ 1.7 = 37.06 = £37.06

Step 3) Choose the COMMON SENSE answer:
Not quite so obvious this time, but since 1.7 US Dollars = £1, you're clearly going to have fewer pounds than you had Dollars (roughly half). In other words, the answer has to be less than 63, so it's £37.06

Conversion Factors

Example 3:

"A popular item at our local Supplies is 'Froggatt's Lumpy Sprout Ketchup' (not available in all areas). The Farmhouse Economy Size is the most popular and weighs 2400 g. How much is this in kg?"

Step 1) Conversion Factor = 1000 (simply because 1 kg = 1000 g)

Step 2) 2400 × 1000 = 2 400 000 kg (Uulp..)
2400 ÷ 1000 = 2.4 kg (that's more like it)

Step 3) So the answer must be that 2400 g = **2.4 kg**

Example 4:

"Toby, my pet elephant, weighs 12 tonnes. How much does he weigh in kg?"

Step 1) Conversion Factor = 1000 (as 1 tonne = 1000 kg)

Step 2) 12 × 1000 = 12 000 kg (looks OK)
12 ÷ 1000 = 0.012 kg (that's a pretty light elephant)

Step 3) So the answer must be that 12 tonnes = **12 000 kg**

Converting Area and Volume Measurements

$1\,m^2 = 100\,cm \times 100\,cm = 10\,000\,cm^2$

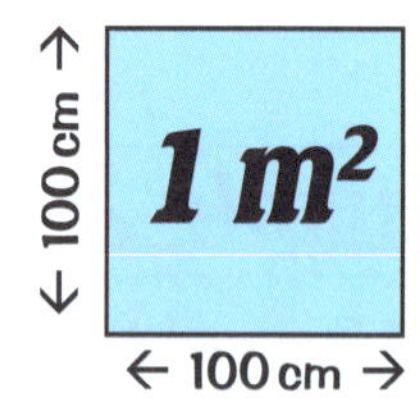

1) To change area measurements from m^2 to cm^2 multiply the area in m^2 by 10 000 (e.g. $3\,m^2 = 30\,000\,cm^2$).

2) To change area measurements from cm^2 to m^2 divide the area in cm^2 by 10 000 (e.g. $45\,000\,cm^2 = 4.5\,m^2$).

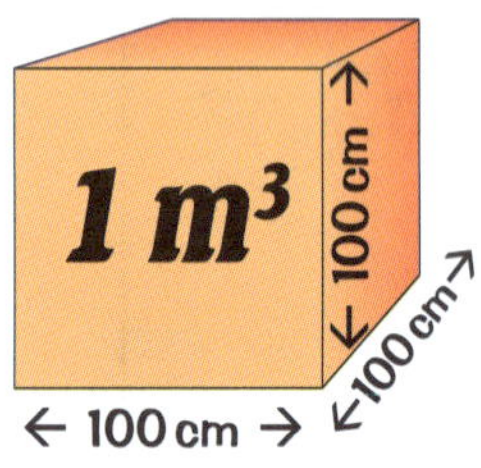

$1\,m^3 = 100\,cm \times 100\,cm \times 100\,cm = 1\,000\,000\,cm^3$

1) To change volume measurements from m^3 to cm^3 multiply the volume in m^3 by 1 000 000 (e.g. $3\,m^3 = 3\,000\,000\,cm^3$).

2) To change volume measurements from cm^3 to m^3 divide the volume in cm^3 by 1 000 000 (e.g. $4\,500\,000\,cm^3 = 4.5\,m^3$).

You don't have the Conversion Factor. I thought it was very karaoke...

What a nice feeling — two lovely pages and just the 3 steps of the Conversion Factor method to learn, and a slightly scarier bit about area and volume conversions. If you're reading this and you've still no idea what the Conversion Factor is — look at all the big colourful boxes on the last two pages.

Metric and Imperial Units

This conversion stuff is easy marks so far — make sure you get them. That sounded a bit like a threat. It wasn't. Honest. I just have poor people skills. Anyway, let's carry on...

Metric Units

1) Length — mm, cm, m, km
2) Area — mm^2, cm^2, m^2, km^2,
3) Volume — mm^3, cm^3, m^3, litres, ml
4) Weight — g, kg, tonnes
5) Speed — km/h, m/s

MEMORISE THESE KEY FACTS:

1 cm = 10 mm	1 tonne = 1000 kg
1 m = 100 cm	1 litre = 1000 ml
1 km = 1000 m	1 litre = 1000 cm^3
1 kg = 1000 g	1 cm^3 = 1 ml

Imperial Units

1) Length — Inches, feet, yards, miles
2) Area — Square inches, square feet, square yards, square miles
3) Volume — Cubic inches, cubic feet, gallons, pints
4) Weight — ounces, pounds, stones, tons
5) Speed — mph

LEARN THESE TOO:

1 Foot = 12 Inches
1 Yard = 3 Feet
1 Gallon = 8 Pints
1 Stone = 14 Pounds (lbs)
1 Pound = 16 Ounces (oz)

Metric-Imperial Conversions

YOU NEED TO LEARN THESE — they DON'T promise to give you these in the Exam and if they're feeling mean (as they often are), they won't.

Approximate Conversions

1 kg = 2.2 lbs	1 gallon = 4.5 litres
1 inch = 2.5 cm	1 mile = 1.6 km
	or 5 miles = 8 km

Imperial units — they're mint...

Learn all the 17 Conversions above, then cover the page and see if you can scribble them all down.

1) a) How many cm is 2 metres? b) How many mm is 6.5 cm?
2) a) How many kg is 2500 g? b) How many litres is 1500 cm^3?
3) A rod is 46 inches long. What is this in feet and inches?
4) a) Roughly how many kg is 110 lb? b) How many cm is 6 feet 3 inches?

Estimating

Everything you need to know about estimating — all on one page. Cashback!*

Estimating to Simplify Calculations

If you need to refresh your memory on rounding, head over to pages 4 & 5.

This is VERY EASY, so long as you don't over-complicate it.

1) ROUND EVERYTHING OFF to nice easy CONVENIENT NUMBERS.
2) Then WORK OUT THE ANSWER using these nice easy numbers — that's it!

EXAMPLE: Estimate the value of $\frac{127.8 + 41.9}{56.5 \times 3.2}$ showing all your working.

ANSWER: $\frac{127.8 + 41.9}{56.5 \times 3.2} \approx \frac{130 + 40}{60 \times 3} = \frac{170}{180} \approx 1$ ("≈" means "roughly equal to")

In the Exam you'll need to show all the steps, to prove you didn't just use a calculator.

Estimating Sizes

This needs a bit of common sense on top of some basic rules to remember.
Have a look at this example:

Example: After a poorly judged business transaction on his way to market, Jack finds himself up a beanstalk and face to knee with a terrifying ogre (shown here).
Assuming Jack is of average height for a strapping young man, estimate the height of the ogre, in m.

1) Firstly, decide on a reasonable height for Jack. About 1.8 m is fine, but you could get away with anywhere from 1.5 to 2 m.
2) Looking at the picture, the ogre is about 3 times Jack's height.
3) Using 1.8 m as Jack's height:

Height of ogre = 3 × 1.8 m = 5.4 m

Areas and Volumes

Draw or imagine a RECTANGLE OR CUBOID of similar size to the object. ROUND OFF all lengths to the NEAREST WHOLE, and work it out — easy.

EXAMPLES: "Estimate the area of this shape and the volume of the bottle:"

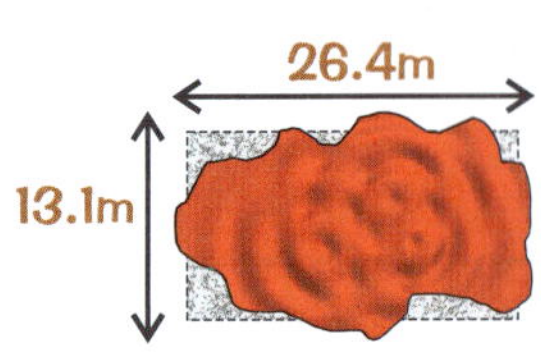

Area ≈ rectangle
26 m × 13 m = 338 m^2
(or without a calculator:
30 × 10 = 300 m^2)

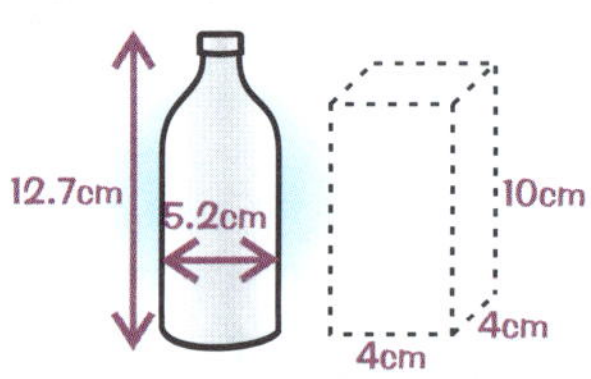

Volume ≈ cuboid
= 4 × 4 × 10
= 160 cm^3

Don't underestimate the importance of eating biscuits when revising...

There's more to estimating than just guesswork. LEARN the rules, then turn over and write them down.

1) Estimate the value of $4.103^2 + (15.86 \div 2.04)$.
2) Estimate the area of Great Britain in square miles, and the volume of a tin of beans in cm^3.

Hint: Land's end to John o'Groats is 605 miles in a straight line.

*Cashback — 'great' / 'bonza' / 'dandy'

Measuring Angles with Protractors

Need to measure an angle? DON'T PANIC, just reach for your protractor and follow this advice...

Using Protractors

The 2 big mistakes that people make with PROTRACTORS:

1) Not putting the 0° line at the start position
2) Reading from the WRONG SCALE.

These scales won't work.

Two Rules for Getting it Right

1) ALWAYS position the protractor with the bottom edge of it along one of the lines as shown here:

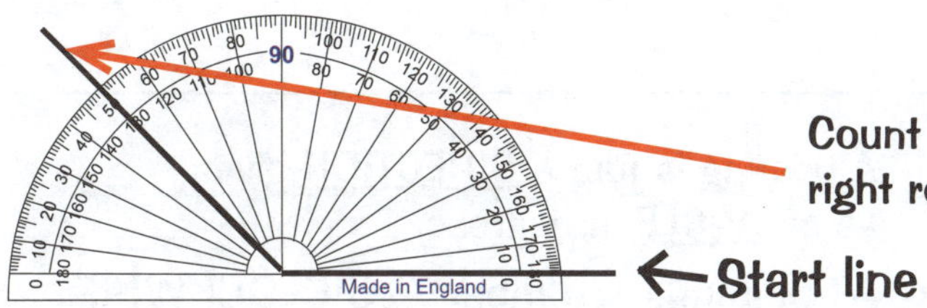

Count in 10° steps from the start line right round to the other one over there.

2) COUNT THE ANGLE IN 10° STEPS from the start line right round to the other one.

> DON'T JUST READ A NUMBER OFF THE SCALE — chances are it'll be the WRONG ONE because there are TWO scales to choose from.
>
> The answer here is 130° — NOT 50° — which you will only get right if you start counting 10°, 20°, 30°, 40° etc. from the start line until you reach the other line. You should also estimate it as a check.

Neither will these.

Three Letter Angle Notation

The best way to say which angle you're talking about in a diagram is by using THREE letters. For example in the diagram, angle ACB = 25°.

1) The MIDDLE LETTER is where the angle is.
2) The OTHER TWO LETTERS tell you WHICH TWO LINES enclose the angle.

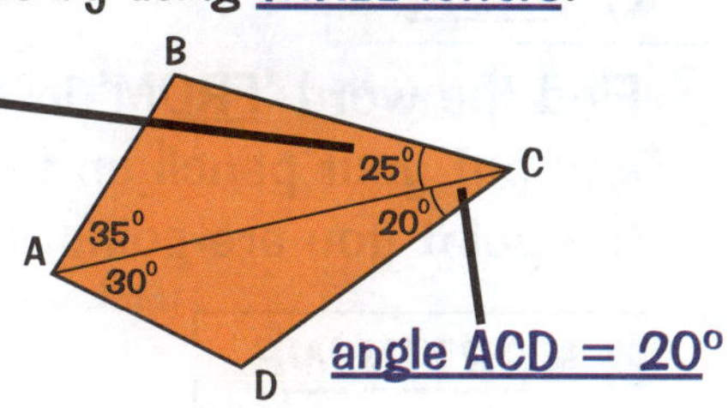

Acute Angles

SHARP POINTY ONES (less than 90°)

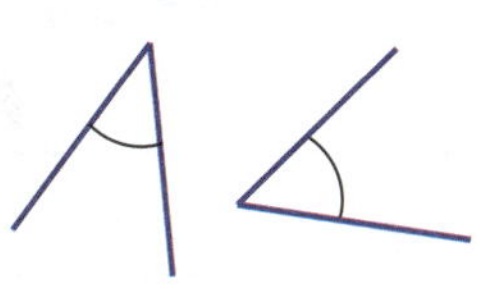

Obtuse Angles

FLATTER-LOOKING ONES (between 90° and 180°)

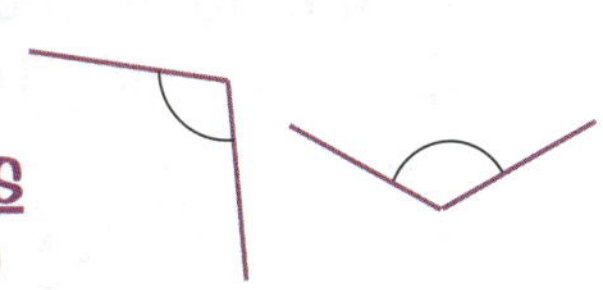

Reflex Angles

ONES THAT BEND BACK ON THEMSELVES (more than 180°)

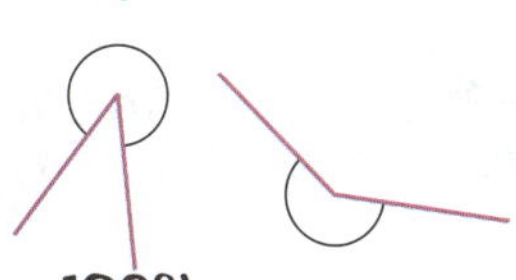

Right Angles

SQUARE CORNERS (exactly 90°)

Roses are red, tulips are plum, if you were an angle you'd be acute one...

Ho, ho, ho — pretty funny aren't I. Anyway, moving on...

1) LEARN 2 rules for using protractors.
2) LEARN what ACUTE, OBTUSE, REFLEX and RIGHT ANGLES are. Draw one example of each.
3) Use a protractor to accurately draw these angles: a) 35° b) 150° c) 80°

Compass Directions and Bearings

You probably already know what the compass is and how it can be really useful. By the end of this page you should have got your bearings about what is meant by bearings too. Ahem.

The Eight Points Of the Compass

Make sure you know all these 8 DIRECTIONS ON THE COMPASS.

For other directions (i.e. not exactly North or South or South-East, etc), you have to use BEARINGS.

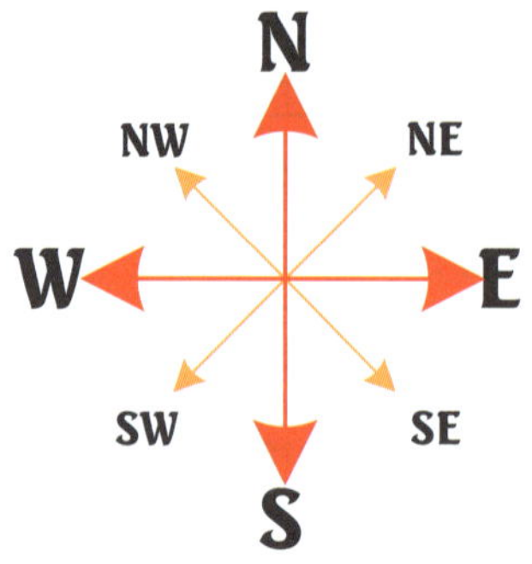

Bearings

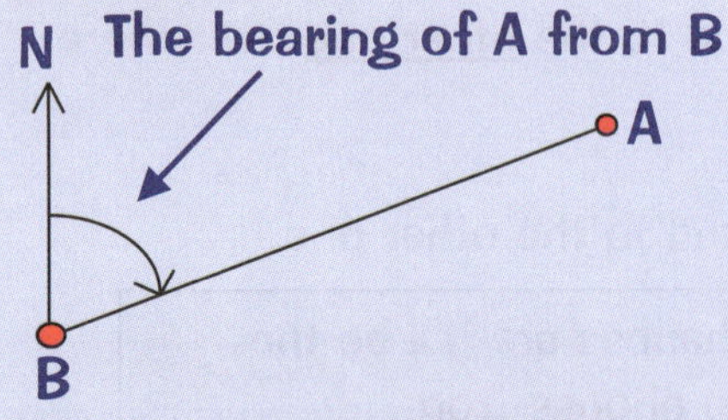

1) A bearing is just a DIRECTION given as an ANGLE in degrees.
2) All bearings are measured CLOCKWISE from the NORTHLINE.
3) All bearings are given as 3 figures: e.g. 060° rather than just 60°, 020° rather than 20° etc.

The 3 Key Words

Only learn this if you want to get bearings RIGHT

1) 'FROM'

Find the word 'FROM' in the question, and put your pencil on the diagram at the point you are going 'from'.

2) NORTHLINE

At the point you are going 'FROM', draw in a NORTHLINE.

3) CLOCKWISE

Now draw in or read off the angle CLOCKWISE from the northline to the line joining the two points. This angle is the BEARING.

EXAMPLE:

Find the bearing of Q from P:

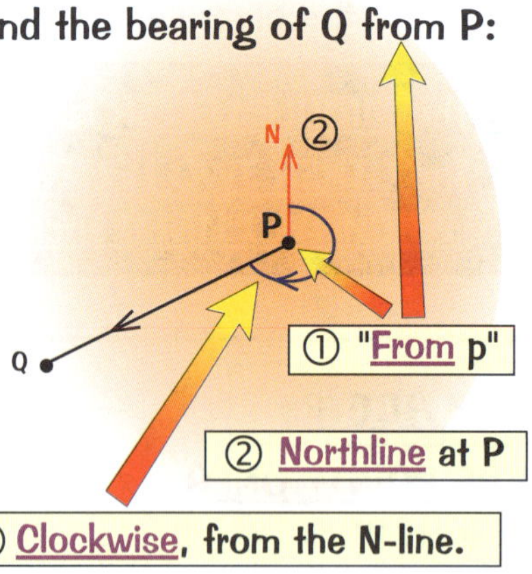

This angle is the bearing of Q from P and is 245°.

Compasses — we'd be lost without them...

Make sure you can draw out the eight points of a compass, and that you know the 3 key words for bearings too. And you'll be reet petite.

1) Draw a blob on a piece of paper to represent home, and then draw 2 lines, one going out in a South-Westerly direction and the other on a bearing of 080°.

Drawing Polygons

That's right, you know you're on to a winner with a page about drawing. Polygons are just shapes with any number of straight sides, and you can draw them perfectly if you follow this simple advice...

Sides and Angles — use a Ruler and Protractor

Example: Construct triangle DEF. DE = 5cm, DF = 3cm, and angle EDF = 40°

ANSWER: First, sketch and label the triangle. Then:

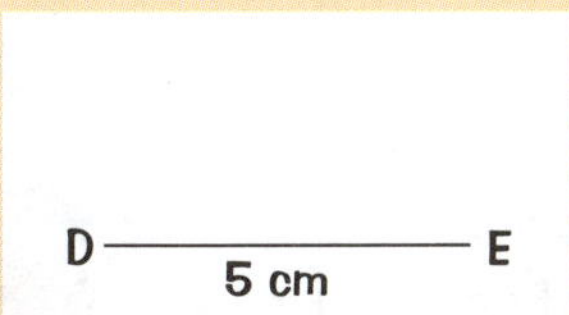

Pick a side for the base line — it doesn't matter which one. I've picked DE. Draw a line 5cm long. Label the ends D and E.

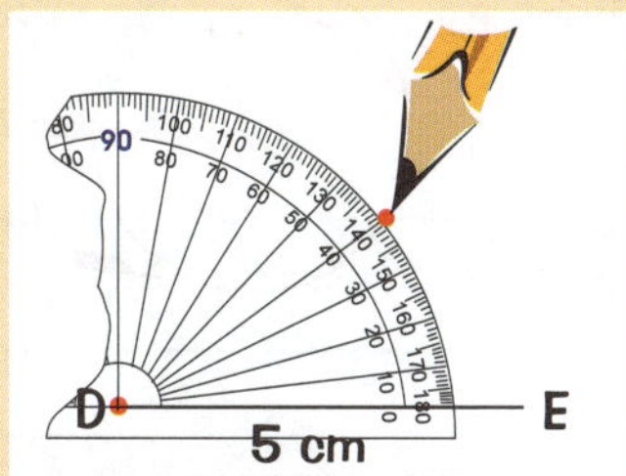

Next you have to draw the angle EDF (the angle at D): Place the centre of the protractor over D, measure 40° and put a dot.

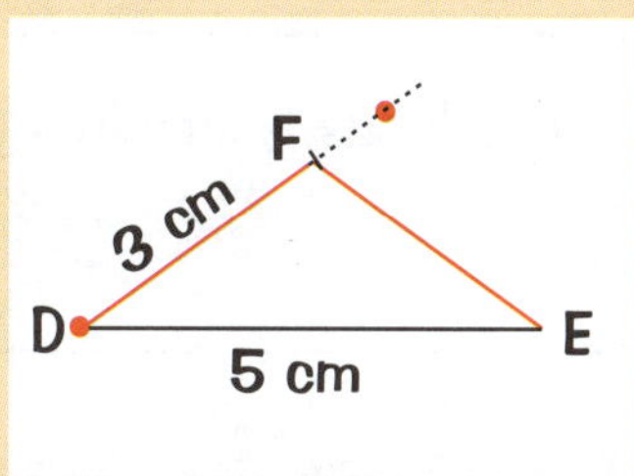

Measure 3 cm towards the dot and label it F. Now you've drawn the two sides and the angle. Just join up F and E to complete the triangle.

Have a look back at page 68 for more about regular polygons

Drawing Regular Polygons

The method is the same for all regular polygons. E.g. "Draw a regular hexagon."

1) Find out the angle in the centre. The total of all the angles is 360°, so divide this by the number of sides: e.g. 360/6 = 60°
2) Draw a circle with a dot on the line.
3) Measure 60° at the centre of the circle from your dot and put another dot.
4) Keep drawing 60° markers 'til you're back at the start.
5) Join the dots with a ruler (and rub out the circle).

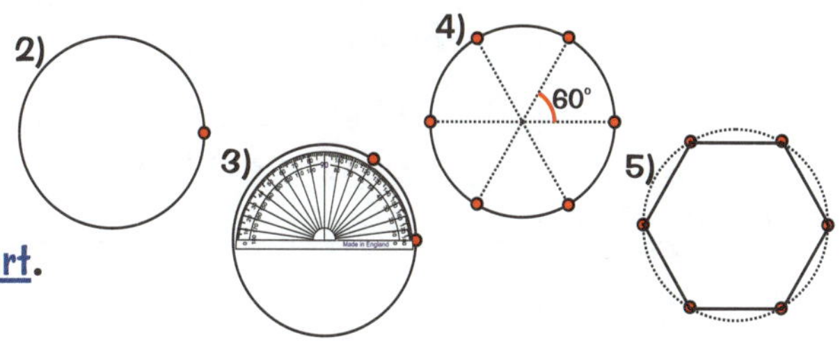

My dad rules at farming — he drives a pro-tractor....

Come along now, you're almost at the end so there's no point hanging around.
Learn this page and have a crack at these little beauties...

1) A triangle ABC has sides AB = 8 cm, AC = 5 cm and angle BAC = 49°. How long is BC to the nearest cm? Construct the triangle to find the answer.
2) Close the book and draw a regular: a) octagon b) pentagon c) hexagon d) triangle

Loci and Constructions

"Loci and constructions — what the monkey is that about" I hear you cry. Well, wonder no more...

Drawing Loci

A **LOCUS** (another ridiculous maths word) is simply:

A LINE that shows all the points which fit in with a given rule

Make sure you learn how to do these PROPERLY using a RULER AND COMPASSES as shown.

The locus of points which are 'A fixed distance from a given point'

This locus is simply a CIRCLE.

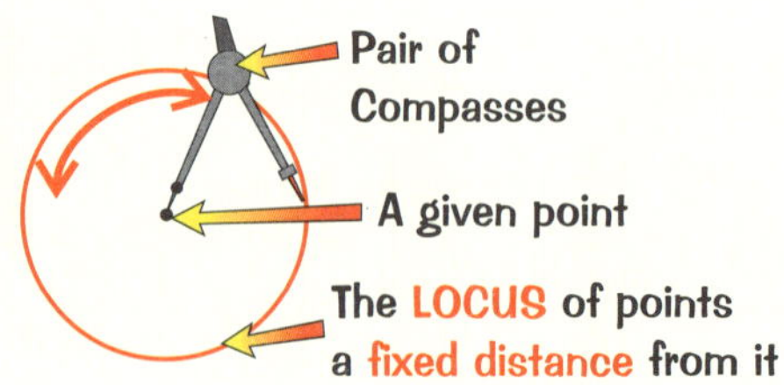

The locus of points which are 'A fixed distance from a given line'

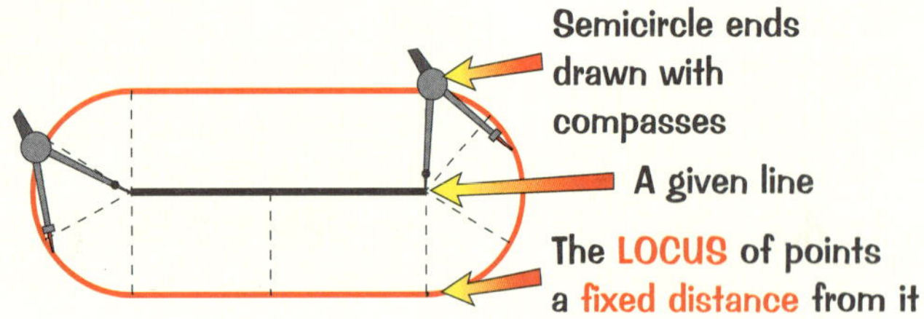

This locus is an OVAL SHAPE

It has straight sides (drawn with a ruler) and ends which are perfect semicircles (drawn with compasses).

The locus of points which are 'Equidistant from two given lines'

1) Keep the compass setting THE SAME while you make all four marks.
2) Make sure you leave your compass marks showing.
3) You get two equal angles — i.e. this LOCUS is actually an ANGLE BISECTOR.

Equidistant just means 'the same distance'.

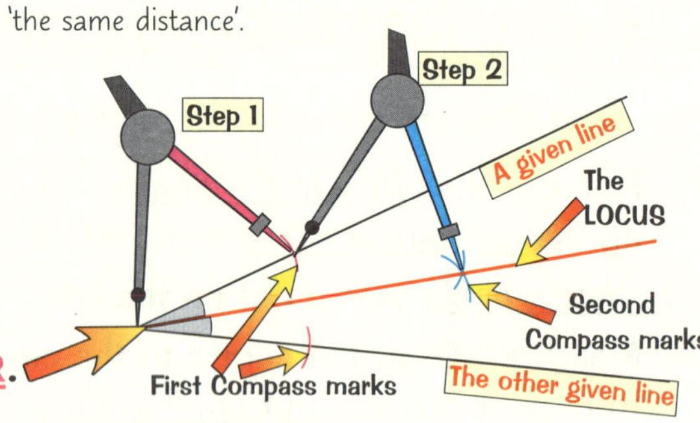

The locus of points which are 'Equidistant from two given POINTS'

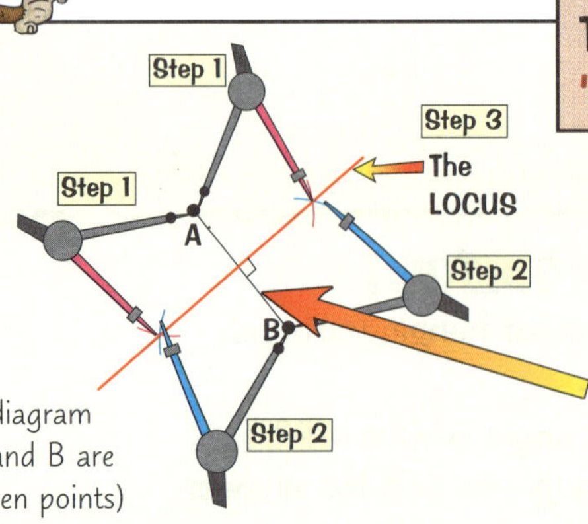

(In the diagram above, A and B are the two given points)

This LOCUS is all the points which are the same distance from A and B.

This time the locus is actually the PERPENDICULAR BISECTOR of the line joining the two points.

Weird, scary, mutant monkey. Run... RUN...

Revision Test for Unit 3 — 2

WHAT YOU'RE SUPPOSED TO DO HERE is put all the methods of the second half of this Unit into practice to answer these questions.

1) What does it mean if two shapes are: a) congruent b) similar?

2) What transformation maps
a) shape A onto shape B?
b) shape C onto shape D?

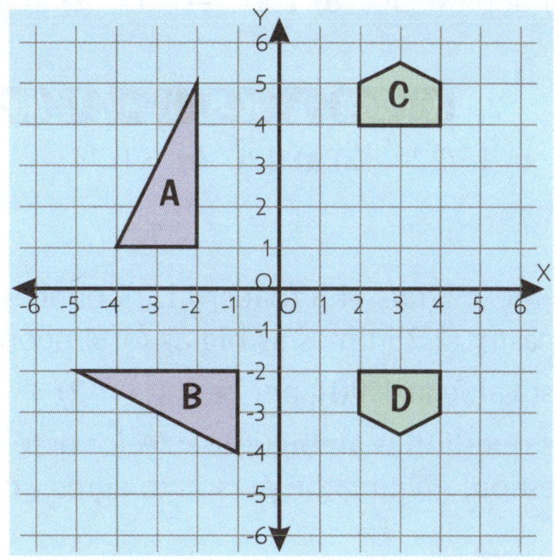

3) Julie wants a picture enlarging for a new frame. Use the diagram to work out the enlargement scale factor Julie needs.

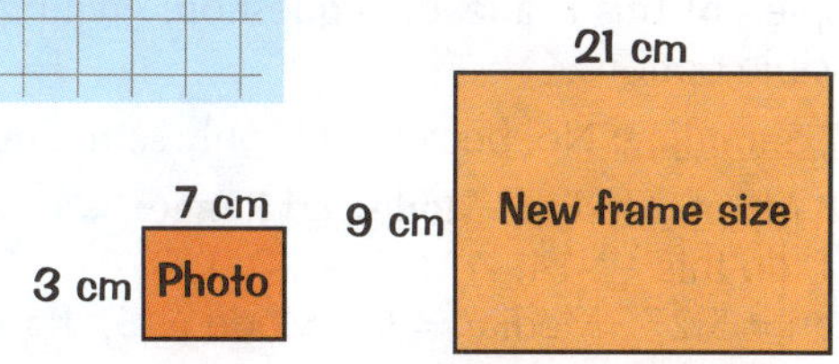

4) A museum has a flight of stairs up to its front door (see diagram). A ramp is to be put over the top of the stairs for wheelchair users. Calculate the length that the ramp would need to be.

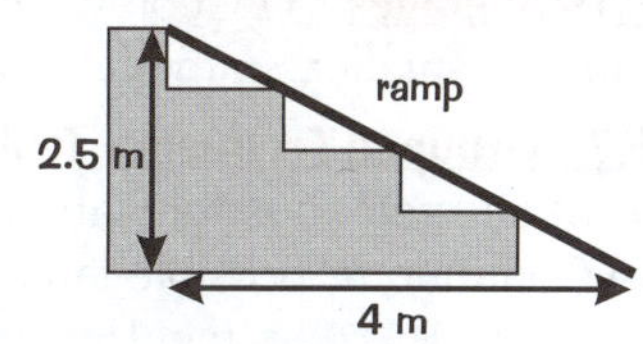

5) Laura is planning a walk in the countryside. The route she has planned is 15 km long. Laura knows that she walks at an average speed of 4 km/h. It gets dark at 5 pm. What is the latest time Laura should set off walking in order to get back before dark?

6) What are the Three Steps of the Conversion Factor Method?

7) Kylie is on holiday in South Africa, where £1 = 12.12 rand. She pays 250 rand to go to a safari park. How much has she spent in pounds?

8) Jane is making pancakes. She has 0.7 litres of milk but needs 550 cm^3 according to the recipe. Does she have enough milk to make the pancakes?

9) 1 mile = 1.6 km. How many miles is 27 km? Give it to the nearest mile.

10) 1 kg = 2.2 lbs. Kevin is filling in a form to join the gym and needs to give his weight in kg. He knows he weighs 10 st 4 lbs. How much is this in kg?

11) Estimate the volume of a mug of black tea in cm^3.

12) A ship sets off from Port P on a bearing of 160°. Show its direction on the drawing opposite:

13) Another ship sets off on a bearing of 240°. Show its direction too:

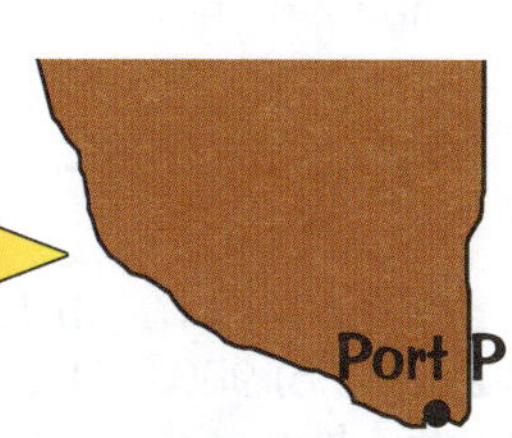

14) Construct a triangle ABC with sides AB = 9 cm, AC = 10 cm and angle BAC = 50°.

15) What is a locus? Describe in detail the four types you should know.

Answers

Unit 1 — Statistics and Number

P.4 Rounding Off: 1) a) 3.2 (1 d.p.), 3.25 (2 d.p.), 3.246 (3 d.p.) **b)** 1.8 (1 d.p.), 1.78 (2 d.p.), 1.780 (3 d.p.) **c)** 2.3 (1 d.p.), 2.31 (2 d.p.), 2.312 (3 d.p.) **d)** 0.5 (1 d.p.), 0.47 (2 d.p.), 0.467 (3 d.p.) **e)** 9.8 (1 d.p.), 9.76 (2 d.p.), 9.765 (3 d.p.) **2) a)** 3 **b)** 5 **c)** 2 **d)** 7 **e)** 3

P.5 Rounding Off: 1) a) 450 **b)** 680 **c)** 50 **d)** 100 **e)** 10 **2) a)** 400 **b)** 500 **c)** 10 **d)** 0.04 **3) a)** 2900 **b)** 500 **c)** 100

P.7 Calculator Buttons: 1) 3/8 **2)** (23.3 + 35.8) ÷ (36 × 26.5)

P.8 Fractions: 1) a) 5/6 **b)** 2/3 **c)** 15/22 **2) a)** 220 **b)** £1.75

P.10 Percentages: 1) £120

P.11 Percentages: 1) 30%

P.12 Ratios: 1) 56p **2)** Recipe for 9: 9 rats, 4.5 toads, 11.25 ounces of "Froggatt's Lumpy Sprout Ketchup", 18 'taties, a very big wodge of pastry (2.25 times as big as for 4 people in fact)

P.13 Ratios: 1) Large size is best value at 1.90 g per penny. **2)** 3 : 8

P.16 Collecting Data: 1) a) This question is ambiguous. "A lot of television" can mean different things to different people. **b)** This is a leading question, inviting the person to agree. **c)** The answers to this question do not cover all possible options.

P.17 Sampling: No, because it is only sampling one school year and music tastes might be different in different year groups.

P.18 Mean, Median, Mode and Range:
First, do this: -14, -12, -5, -5, 0, 1, 3, 6, 7, 8, 10, 14, 18, 23, 25
Mean = 5.27, Median = 6, Mode = -5, Range = 39

P.19 Averages: 1) Geoff would choose variety X because the mean number of potatoes per plant is higher. Also, the range is smaller, so the number of potatoes produced by each plant is more consistent than for variety Y.

P.21 Grouped Frequency Tables:
1) a) Discrete, possible table: Columns "Shoe Size" and "frequency", Shoe size intervals: 3, 4, 5, 6, 7, 8, 9, 10 , 11, 12.
b) Continuous, possible table: Columns "Height (cm)" and "frequency",
intervals: $140 \le h < 150$, $150 \le h < 160$, $160 \le h < 170$, etc.

P.22 Grouped Frequency Tables: (See table.)
1) Mean = 17.4 cm **2)** Modal Class = $17.5 \le L < 18.5$
3) Median is in the group $17.5 \le L < 18.5$

Length L (cm)	$15.5 \le L < 16.5$	$16.5 \le L < 17.5$	$17.5 \le L < 18.5$	$18.5 \le L < 19.5$	Totals
Frequency	12	18	23	8	61
Mid-Interval Value	16	17	18	19	—
Freq × MIV	192	306	414	152	1064

P.24 Tables, Charts and Graphs:

1)
0 | 3 6 7
1 | 1 2 3 4 6 7 9
2 | 0 2 4 6 6

2) Median = 16, Range = 23

P.26 Bar Charts and Comparing Data: 1) See bar chart.

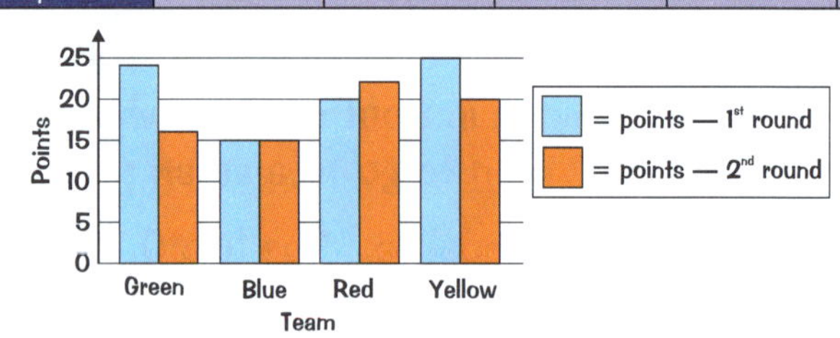

P.27 Pie Charts: See pie chart to the right.

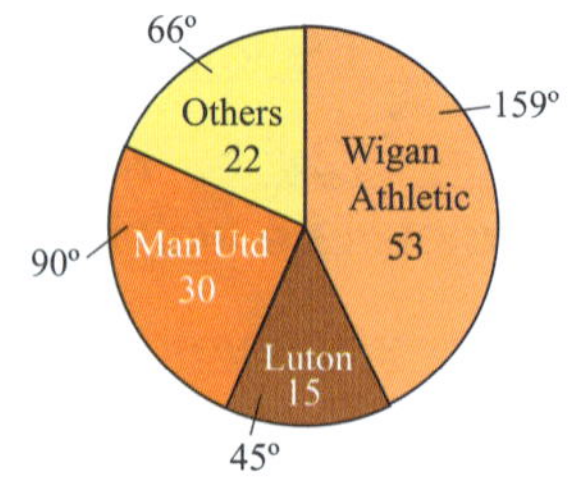

P.28 Probability: 1) a) 1/13 **b)** 6/13 **c)** 3/26

P.29 Probability: 1) 1/5 or 20% or 0.2

P.30 Relative Frequency:
1) Landing on red: 0.43, landing on blue: 0.24, landing on green: 0.33

Revision Test for Unit 1 — 1

1) a) 1 **b)** 3 **c)** 16 **d)** 12 **2) a)** 5.3 **b)** 3.5 **c)** 6.2 **3) a)** 5.33 **b)** 3.46 **c)** 6.16 **4) a)** 5.327 **b)** 3.464 **c)** 6.159
5) a) 250 **b)** 900 **6) a)** 2 **b)** 2 **c)** 1 **d)** 3 **e)** 2 **7)** 4/5 **8) a)** 320 **b)** £60 **c)** 195 **9)** £58.75
10) No, it would cost £36 **11)** Golden Rule: "DIVIDE FOR ONE THEN TIMES FOR ALL" **12)** £1.41
13) Golden Rule: "DIVIDE BY THE PRICE IN PENCE" **14)** Biggest is best buy at 4.8 g per penny
15) a) 1 : 9 **b)** 5 : 8 **c)** 2 : 3

Answers

Revision Test for Unit 1 — 2

1) Any three from: make sure your question is relevant; make your question clear, brief and easy to understand; allow for all possible answers; make sure your question isn't leading or biased; make sure your question isn't ambiguous; make sure your question isn't vague.

2) First put in order: 2, 3, 4, 6, 7, 7, 12, 15 **a)** 7 **b)** 6.5 **c)** 7 **d)** 13 3) Values of w from 50 to 60 **including** 50, but **not including** 60. 60 would go in the next group up. 4) Mean ≈ ((75 × 15) + (105 × 60) + (135 × 351) + (165 × 285) + (195 × 206) + (225 × 83)) ÷ 1000 = 160 680 ÷ 1000 = 161 min

5) Pictogram, 35 angry customers 6) **a)** Both things increase or decrease together and they're closely related. (The more cheese you eat, the more you tend to have nightmares.)
b) No, the graph shows very weak correlation.

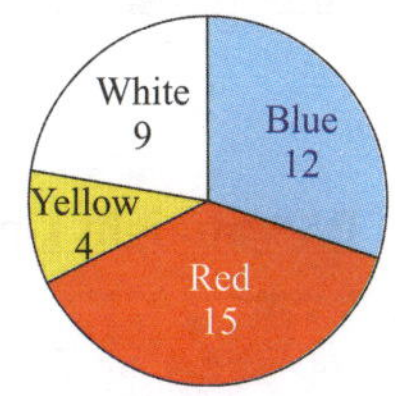

7) **a)** Angles are: Blue 108°, Red 135°, Yellow 36°, White 81°. **b)** See pie chart.

8) H-H, H-T, T-H, T-T ¼ 9) H-1, H-2, H-3, H-4, H-5, H-6, T-1, T-2, T-3, T-4, T-5, T-6 1/12

10) 0.8 11) **a)** 53.3% **b)** Add one more black ball, or take away one red or one green ball.

Unit 2 — Number and Algebra

P.32 Big Numbers:

1) **a)** One million, two hundred and thirty-four thousand, five hundred and thirty-one
b) Twenty-three thousand, four hundred and fifty-six **c)** Two thousand, four hundred and fifteen
d) Three thousand, four hundred and two **e)** Two hundred and three thousand, four hundred and twelve

2) 56,421 **3)** 9, 23, 87, 345, 493, 1029, 3004 **4)** 0.008, 0.09, 0.1, 0.2, 0.307, 0.37

P.33 Multiplying by 10, 100, etc.: **1) a)** 1230 **b)** 3450 **c)** 9650 **2 a)** 48 **b)** 450 **c)** 180 000

P.34 Dividing by 10, 100, etc.: **1) a)** 0.245 **b)** 6.542 **c)** 0.00308 **2 a)** 1.6 **b)** 12 **c)** 5

P.36 Dividing Without a Calculator: **1)** 336 **2)** 616 **3)** 832 **4)** 12 **5)** 121 **6)** 21 **7)** 179.2 **8)** 6.12 **9)** 56.1 **10)** 56 **11)** 46 **12)** 12

P.37 Prime Numbers: **1)** 2, 3, 5, 7, 11, 13, 17, 19, 23, 29, 31, 37, 41, 43, 47 **2)** 97, 101, 103, 107, 109

P.38 Multiples, Factors, Prime Factors:

1) 7, 14, 21, 28, 35, 42, 49, 56, 63, 70 and 6, 12, 18, 24, 30, 36, 42, 48, 54, 60
2) 1, 2, 3, 4, 6, 9, 12, 18, 36 and 1, 2, 3, 4, 6, 7, 12, 14, 21, 28, 42, 84
3) a) 990 = 2 × 3 × 3 × 5 × 11 **b)** 160 = 2 × 2 × 2 × 2 × 2 × 5

P.39 LCM and HCF: **1)** 8, 16, 24, 32, 40, 48, 56, 64, 72, 80 and 9, 18, 27, 36, 45, 54, 63, 72, 81, 90 LCM = 72
2) 1, 2, 4, 7, 8, 14, 28, 56 and 1, 2, 4, 8, 13, 26, 52, 104 HCF = 8 **3)** 63 **4)** 12

P.40 Special Number Sequences:

1) EVENS: 2, 4, 6, 8, 10, 12, 14, 16, 18, 20, 22, 24, 26, 28, 30
ODDS: 1, 3, 5, 7, 9, 11, 13, 15, 17, 19, 21, 23, 25, 27, 29
SQUARES: 1, 4, 9, 16, 25, 36, 49, 64, 81, 100, 121, 144, 169, 196, 225
CUBES: 1, 8, 27, 64, 125, 216, 343, 512, 729, 1000, 1331, 1728, 2197, 2744, 3375
POWERS OF 2: 2, 4, 8, 16, 32, 64, 128, 256, 512, 1024, 2048, 4096, 8192, 16 384, 32 768
POWERS OF 10: 10, 100, 1000, 10 000, 100 000, 1 000 000, 10 000 000, 100 000 000, 1 000 000 000, 10 000 000 000, 100 000 000 000, 1 000 000 000 000, 10 000 000 000 000, 100 000 000 000 000, 1 000 000 000 000 000 hmm...

2) **a)** 56, 134, 156, 36, 64 **b)** 23, 45, 81, 25, 97, 125, 1 **c)** 81, 25, 36, 1, 64 **d)** 125, 1, 64 **e)** 1, 64

P.41 Powers: **1) a)** 3^8 **b)** 4 **c)** 8^{12} **d)** 1 **e)** 7^6 **2) a)** 5^{12} **b)** 36 or 6^2 **c)** 2^5

P.42 Square Roots and Cube Roots: **1) a)** 9 **b)** 3, other val in **a)** is -9 **2) a)** g = 6 or -6 **b)** b = 4 **c)** r = 3 or -3

P.43 Fractions, Decimals and Percentages:

1) a) 6/10 = 3/5 **b)** 2/100 = 1/50 **c)** 77/100 **d)** 555/1000 = 111/200 **e)** 56/10 = 28/5 or 5 3/5

P.44 Fractions: **1) a)** 5/32 **b)** 32/35 **c)** 23/20 = 1 3/20 **d)** 1/40 **e)** 167/27 = 6 5/27 **2)** 3/5, 2/3, 11/15

P.45 Ratios: **1)** 24 **2)** £1000 : £1400

P.47 Negative Numbers and Letters: **1) a)** +12 (Rule 1) **b)** -6 (Rule 1/Rule 2)
c) x (Rule 2, then Rule 1) **d)** -3 (Rule 1) **2) a)** +18 **b)** -216 **c)** 2 **d)** -27 **e)** -336

P.48 Algebra: **1) a)** $4x + y - 4$ **b)** $9x + 5xy - 5$ **c)** $5x + 3x^2 + 5y^2$ **d)** $6y - 4xy$

P.49 Algebra: **1) a)** $2x - 4$ **b)** $5x + x^2$ **c)** $y^2 + xy$ **d)** $6xy - 18y$ **2) a)** $5x(y + 3)$ **b)** $a(5 - 7b)$ **c)** $6y(2x + 1 - 6y)$

P.50 Formulas: **1)** $y = 5x - 3$ **2)** 25

Answers

P.51 Solving Equations: 1) $x = 8$ **2)** $x = 7$

P.52 Solving Equations: 1) a) $x = 4$ **b)** $q = 32$ **c)** $y = -2$ **2)** $b = \frac{1}{2}a + 3$

P.53 Inequalities: 1) $x \geqslant -2$ **2)** $x \geqslant -4, x < 2$, $x = -4, -3, -2, -1, 0, 1$

P.55 Number Patterns and Sequences: 1) a) 20, 27 "Add one extra each time" **b)** 2000 20 000 "Multiply the previous term by 10" **c)** 4, 2 "Divide the previous term by 2" **2)** $2n + 5$

P.56 Coordinates: A(4, 5) B(6, 0) C(5, -5) D(0, -3) E(-5, -2) F(-4, 0) G(-3, 3) H(0, 5)

P.58 Straight Line Graphs: 1) And **2)**

x	-4	-2	-1	0	1	2	4
y	-6	-4	-3	-2	-1	0	2

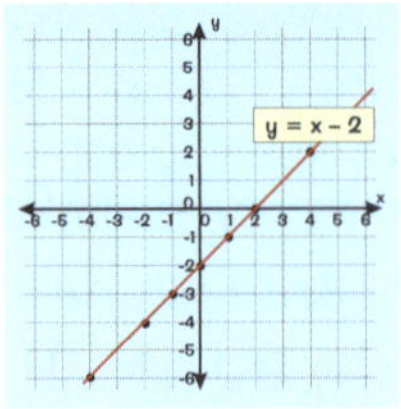

P.59 Straight Line Graphs — Gradients: Gradient = -1.5

Revision Test for Unit 2 — 1

1) Twenty-one million, three hundred and six thousand, five hundred and fifteen.
2) 2, 23, 45, 123, 132, 789, 2200, 6534 3) £120 4) 32p 5) 50p
6) 41, 43, 47, 53, 59 7) Multiples are a number's times table; 10, 20, 30, 40, 50, 60; 4, 8, 12, 16, 20, 24; 8) Factors are what divide into a number ; 1, 2, 3, 5, 6, 10, 15, 30
9) **a)** $210 = 2 \times 3 \times 5 \times 7$ **b)** $1050 = 2 \times 3 \times 5 \times 5 \times 7$ 10) 14 11) 40 12) See P.40
13) **a)** 6 **b)** 1 **c)** 64 **d)** 1 **e)** 32 **f)** 64 14) Approximately 12^4 15) **a)** +15, –15 **b)** 6
16) 645/1000 = 129/200 17) **a)** 88/30 = 44/15 = $2\frac{14}{15}$ **b)** 75/48 = 25/16 = $1\frac{9}{16}$ **c)** 23/8 = $2\frac{7}{8}$ **d)** 11/21
18) See P.45 19) 240

Revision Test for Unit 2 — 2

1) **a)** +6 **b)** –32 **c)** –3 **d)** +5 2) $5x$ 3) $12g + 20h - 4$ 4) $2x(1 + 3y)$ 5) $y = 2x + 4$
6) **a)** $F = \frac{9}{5}C + 32$ **b)** $F = \frac{9}{5}(30) + 32 = 86$ °F 7) **a)** $x = 2$ **b)** $x = 3$ **c)** $x = -7$
8) **a)** $y = 6 - x$ **b)** $y = (x - 11) / 2$ **c)** $y = 3(7x + 3)$ or $y = 21x + 9$ 9) **a)** $x < 5$ **b)** $x < -1$ **c)** $x > -5$ **d)** $x > 10$
10) **a)** 19, 23; add 4 to the previous term **b)** 6, 3; subtract one less each time **c)** 48, 96; double the previous term **d)** 150, 75; halve the previous term **e)** 31, 50; add the previous two terms
11) **a)** nth term = $6n - 7$ **b)** 10th term = $6 \times 10 - 7 = 53$ 12) See right
13) See right and below

x	-5	-3	-1	0	1	2	4	6
y	-2	0	2	3	4	5	7	9

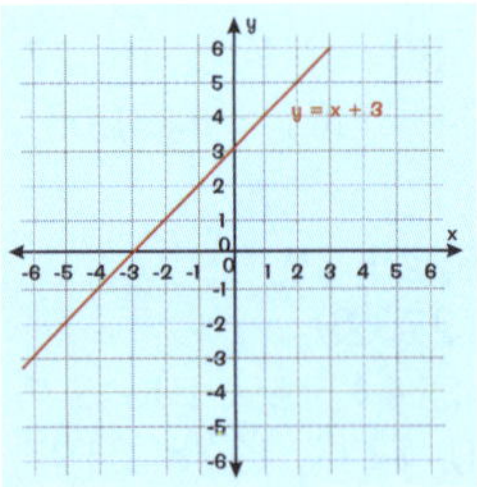

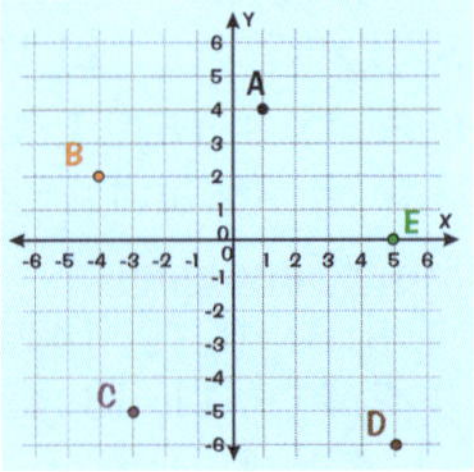

14) Gradient m = 3
15) **a)** Wandering away from home. **b)** Stood still chomping grass. **c)** 2½ km

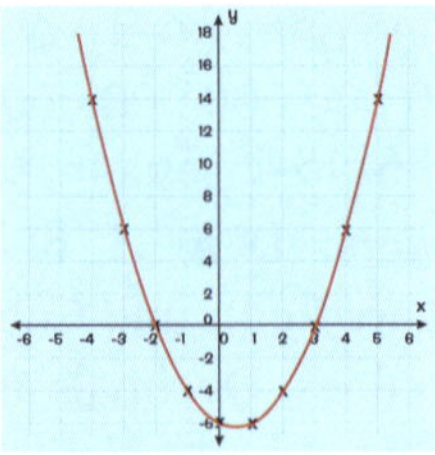

Unit 3 — Geometry and Algebra

P.63 Trial and Improvement: $x = 1.6$

P.64 Quadratic Graphs:
See graph to the right. $x = -2$ and $x = 3$.

P.65 Symmetry:
H : 2 lines of symmetry, Rot^nl symmetry Order 2,
T : 1 line of symmetry, No Rotational symmetry,
E : 1 line of symmetry, No Rotational symmetry,
S : 0 lines of symmetry, Rot^nl symmetry Order 2
Z : 0 lines of symmetry, Rot^nl symmetry Order 2,
N : 0 lines of symmetry, Rot^nl symmetry Order 2,
X : 4 lines of symmetry, Rot^nl symmetry Order 4,

P.68 Regular Polygons:
1) A regular polygon is a many-sided shape where all the sides and angles are the same **2)** Equilateral triangle, square, regular pentagon, regular hexagon, regular heptagon, regular octagon **3)** Ext. angle = 72°, Int. angle = 108° **4)** Ext angle = 30°, Int. angle = 150°

Answers

P.69 Perimeters: 2) 42 cm

P.71 Areas: 1) 12 cm^2 **2)** 12 m^2 **3)** 21 m^2 **4)** 78 cm^2

P.73 Circle Questions: 1) Area = 154 cm^2, circumference = 44 cm **2)** 38.2 turns

P.74 Volume: a) V = 148.5 cm^3 **b)** V = 0.700 m^3 or 700 000 cm^3

P.75 Solids and Nets: 1) 128.8 cm^2 **2)** 294 cm^2 **3)** 180.6 cm^2 (to 1 d.p.) **4)** 96 cm^2

P.77 Parallel Lines: See right

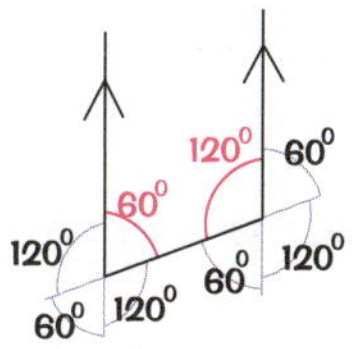

P.79 Congruence and Similarity: 1) a) i, ii and iv are similar. **b)** i and ii are congruent.

P.81 The Four Transformations: 1) A → B Rotation 90° clockwise about the origin. B → C reflection in the line y = x. C → A reflection in the y-axis. A → D translation of $\binom{-9}{-7}$

P.83 Enlargements: 1) 2430 cm^2 **2)** 1875 cm^3

P.84 Pythagoras' Theorem: 1) BC = 8 m,
2) 5 m, 12 m, 13 m is a right angled triangle because $a^2 + b^2 = h^2$ works.

P.87 Maps and Map Scales: 1) 1400 m **2)** 2½ cm

P.91 Metric and Imperial Units: 1) a) 200 cm **b)** 65 mm **2) a)** 2.5 kg **b)** 1.5 litres **3)** 3 feet 10 inches
4) a) 50 kg **b)** 187.5 cm

P.92 Estimating: 1) Approx $4^2 + (16 \div 2) = 24$
2) GB: Approx 600 miles × 150 miles = 90,000 sq. miles.
Bean Tin: Approx 7 cm × 7 cm × 10 cm high = 490 cm^3.

P.93 Measuring Angles with Protractors: 1) and 2) See P.93 **3)**

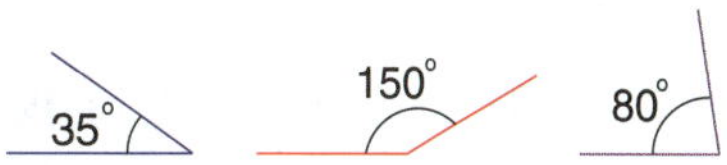

P.94 Compass Directions and Bearings: 1) See right.

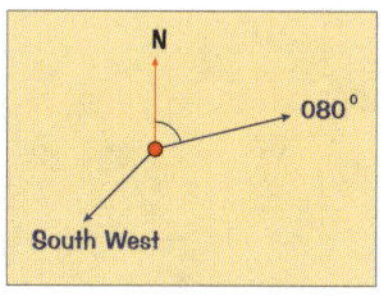

P.95 Drawing Polygons: 1) BC = 6cm **2)** See P.68.

Revision Test for Unit 3 — 1

1) x = 5.5 to 1 d.p. 2) See graph to the right.

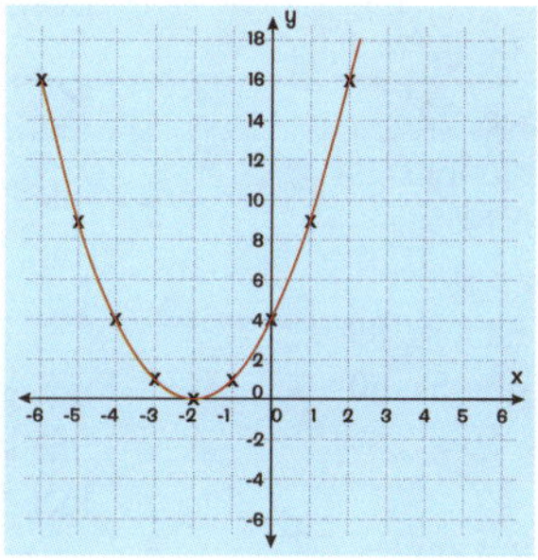

3) Line symmetry and rotational symmetry.

4) a) b) c) d)

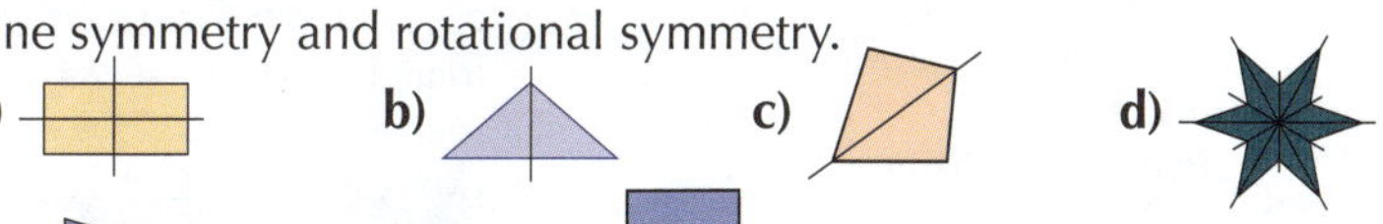

5) a) b) c)

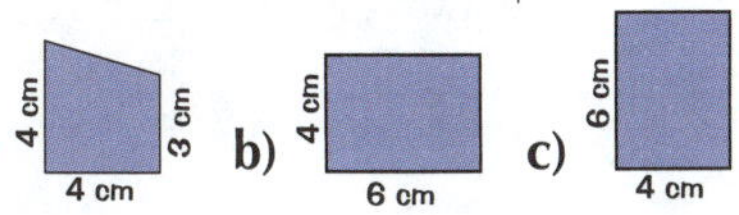

6) 45° and 135° 7) 28 m 8) £1008

9) A = π × r^2 C = π × D 10) a) π is approximately 3.14 b) 6 m 11) See P.72

12) 37.7 cm 13) 3 14) E.g. a cuboid measuring 3 cm by 2 cm by 5 cm

15) See P.75 16) X = 110°, Y = 40°

17) Z = Alternate, C = supplementary, U = supplementary and F = corresponding. See P.77 for sketches.

Revision Test for Unit 3 — 2

1) See P.79. 2) a) 90° anticlockwise rotation about the origin b) Reflection in the line y = 1 3) 3
4) 4.72 m 5) 1:15 pm 6) See P.89 7) £20.63 8) Yes 9) 17 miles 10) 65.45 kg
11) Around 400 cm^3 12) and 13) See right.
15) See P.96.

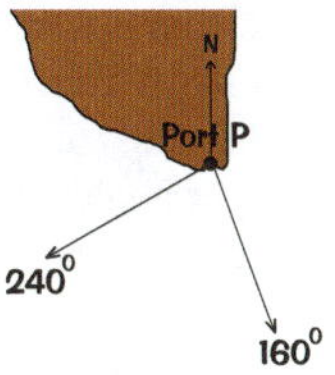

Index